LOCAL
MATHEMATICS FOR
LOCAL PHYSICS
From Number Scaling to
Gauge Theory and Cosmology

LOCAL MATHEMATICS FOR LOCAL PHYSICS

From Number Scaling to Gauge Theory and Cosmology

Paul Benioff

Edited by

Marek Czachor

Gdańsk University of Technology, Poland

Foreword by

Seth Lloyd

Massachusetts Institute of Technology, USA

 World Scientific

NEW JERSEY · LONDON · SINGAPORE · BEIJING · SHANGHAI · HONG KONG · TAIPEI · CHENNAI · TOKYO

Published by

World Scientific Publishing Europe Ltd.

57 Shelton Street, Covent Garden, London WC2H 9HE

Head office: 5 Toh Tuck Link, Singapore 596224

USA office: 27 Warren Street, Suite 401-402, Hackensack, NJ 07601

Library of Congress Cataloging-in-Publication Data
Names: Benioff, P. A., author. | Czachor, Marek, editor.
Title: Local mathematics for local physics : from number scaling to gauge theory and cosmology /
    Paul Benioff ; edited by Marek Czachor, Gdańsk University of Technology, Poland.
Description: New Jersey : World Scientific, [2024] | Includes bibliographical references and index.
Identifiers: LCCN 2023032815 | ISBN 9781800614963 (hardcover) |
    ISBN 9781800614970 (ebook for institutions) | ISBN 9781800614987 (ebook for individuals)
Subjects: LCSH: Mathematical physics. | Gauge fields (Physics)--Mathematics. |
    Vector analysis. | Fiber spaces (Mathematics)
Classification: LCC QC20.6 .B46 2024 | DDC 530.15--dc23/eng/20231023
LC record available at https://lccn.loc.gov/2023032815

British Library Cataloguing-in-Publication Data
A catalogue record for this book is available from the British Library.

For any available supplementary material, please visit
https://www.worldscientific.com/worldscibooks/10.1142/Q0442#t=suppl

Desk Editors: Soundararajan Raghuraman/Rosie Williamson/Shi Ying Koe

Typeset by Stallion Press
Email: enquiries@stallionpress.com

Printed in Singapore

In Memory of Paul Benioff

Paul Benioff poured all his energies into completing this book over the last several years of his life. He devoted his career and most of his waking hours (and likely some of his dreams) to developing concepts that would pave the way toward a more coherent theory of math and physics. This book reflects the culmination of the concepts on this journey that he developed during the later stages of his career.

Paul's dedication and persistence in tackling some of the most vexing issues at the frontier of physics and math was a constant source of inspiration for us and all who knew him. While he was a humble scientist, he constantly pushed himself to formulate and test innovative theories and proofs on highly complex and poorly understood topics. He would frequently work from a blank piece of paper in developing new concepts buttressed with elaborate mathematical formulas.

His work developing one of the first theoretical models of quantum computing garnered some well-deserved recognition from his peers. But he otherwise toiled largely in anonymity and by himself, dependent on his deep curiosity and enduring drive to carry him forward. We all so greatly admired his ability to persevere in seeking answers to deep issues at the very limits of our understanding.

Along with being a prolific scientist, Paul was a caring husband and father who loved nothing more than a gathering of family and friends. He also was passionate about gardening and exploring the mysteries of nature, especially hiking in the California Sierras.

We dedicate this book to Paul and the incredible richness, love, and constant intellectual insights he brought to our lives and that of many others.

We would like to express our deepest gratitude to Marek Czachor who spent many hours polishing and editing the draft which was prepared by Paul before his death in 2022. We also greatly appreciate the thorough work of World Scientific Publishing in editing and publishing this document, as well as to Seth Lloyd for writing the foreword.

With much gratitude and love,
Hanna (wife), Ron (son),
Leora (daughter), and Ilana (daughter)

Foreword

Paul Benioff was a brilliant scientist who delighted in thinking about problems in ways that had occurred to no one else. He is best known as the first person to propose quantum computers: beginning in 1980, two years before Feynman's work on universal quantum simulators, Benioff leveraged recent discoveries of reversible logic to construct both unitary and Hamiltonian dynamical models for quantum computation. By 1985, when David Deutsch introduced the concept of quantum parallelism — the basis for quantum computational speed-ups over classical computers — Benioff's papers had already shown how quantum computation could be embedded in physical systems.

Benioff's great works rose out of his incessant asking of fundamental questions. Do the laws of quantum mechanics support the construction of computers in which each atom or electron registers a bit of information? Benioff's answer is yes. This book asks a particularly primeval question: Could the concept of number depend on where one is in space?

Because the laws of physics are based on spatially local interactions, to compare a mathematical quantity at point A with the analogous concept at point B, information has to be sent from A to B, and vice versa. To construct his theory of general relativity, Einstein had already introduced the concept that quantities such as direction and length required a so-called "connection" to compare those quantities at different points. Elementary particle physics is fundamentally based on a similar concept: to compare physical quantities such as the electric or magnetic fields from point to point, one requires a special type of connection called a gauge field.

In this, his last work, Benioff assembles his attempts to try to answer a fundamental question that he had begun to ask decades before: Does the very concept of a number, "4", for example, vary from point to point, and do we require a way to compare ostensibly identical numbers at different points? That is, does Einstein's concept of a connection to compare physical directions and lengths at different points generalize to a connection for comparing abstract mathematical objects themselves? Benioff's ambitious answer is yes: he proposes a quantity called a "value field" that allows one to compare the abstract concept of a number from point to point. He then explores the possible implications of such a field for problems in quantum physics and cosmology, putting forward several intriguing conjectures about the nature of physical reality.

It is too early to tell whether — like his work on quantum computation — Benioff's novel insight will give rise to a fundamental way of reconceiving mathematics and physics. Benioff was a revolutionary thinker, and not all revolutions work out! The loving care for explaining hard concepts which he lavished on this book, sensitively edited by Marek Czachor, makes it essential reading for any thinker who wishes to be taken on a tour of the cosmos by one of its most thoughtful inhabitants.

Seth Lloyd
Massachusetts Institute of Technology

Contents

Part I
Local Mathematics and Value Fields

Chapter 1

Introduction

1.1 Background

Mathematics plays an essential role in much of science. Theories in physics and in other areas of science have an essential mathematical component. Theory predictions are often obtained from solving various mathematical equations. Support or refutation of the theory is based on agreement or disagreement between theory and experiment. This use of mathematics to construct theories in physics has been enormously successful. One needs only to witness the great success and advancement of our understanding of the world around us.

Nevertheless, one does wonder why mathematics is so successful. This question was the basis of the well-known paper by Wigner [1] on "The unreasonable effectiveness of mathematics in the natural sciences". This question is especially relevant if one takes the view that mathematical entities have an ideal existence outside of or independent of space and time, and physical systems move in and affect space and time. Why should systems or entities existing outside space and time be so useful in describing systems moving in and affecting space and time?

The large literature [2–5] stimulated by Wigner's paper and the great success of mathematics in the development of physical theories lead one to think that there must be a very deep and close relationship between mathematics and physics at some foundational level. This has led to the concept of a coherent theory of physics and mathematics together [6,7]. One would expect such a theory to treat physics and mathematics together as one coherent whole rather

3

than as two separate types of systems. Such a theory must at least work with some idea of what mathematical systems are and what mathematics describes. Without this it seems hopeless to attempt to understand the deep relation between physical and mathematical systems and their properties.

It is not the purpose here to go into detail on the large literature on the nature and meaning of mathematics and mathematical objects [8,9] and on the relation between mathematics and physics [10–12]. Nevertheless, one needs to have some idea of what mathematics is about.

Here it is assumed that mathematics consists of the study of mathematical systems of many different types and of the relations between the systems. Systems of different types are considered to be represented by structures [13,14] relevant for the system type. A structure consists of a base set, a few basic operations, none or a few basic relations, and none or a few constants. The structure must satisfy a set of axioms that are relevant for the system type. In addition, there are relations or maps between structures of different types. The description and properties of these maps are often included in the axioms.

The use of structures to represent mathematical systems is in accordance with the description in mathematical logic of structures as semantic models of systems in which the axioms are true. These structures have meaning or value as distinct from the purely formal syntactic structures as meaningless strings of symbols. The fact that structures are meaningful is relevant to this work because it brings the conscious observer into the description [15]. Meaning or value is an observer-related concept.

A basic aspect of the use of mathematics in physics and other sciences is that mathematical systems are assumed to be outside of space or time. The mathematical systems used to derive theory predictions have no space or time location associated with them. They may be interpreted to describe a system at a particular location but the mathematical object itself is not located in space or time. As an example, a quantum wave packet state of a particle may describe a particle with a high probability of being in a particular region of space, but the wave packet itself, as a vector in a Hilbert space, is not associated with any space or time location. Mathematical systems or elements not located at a space or time point or within a space or time regions are referred to here as systems or elements of a global mathematics.

Gauge theories provide an exception to this assumption. In these theories, a separate vector space is associated with each point on an underlying space–time [16,17]. Each vector space, as a mathematical system or structure, is a local mathematical structure.

A matter field is a map from space–time to the local vector spaces such that for each space–time point the field value, as a vector, is a member of the local vector space at the point. One sees from this that each value of the field is local in that it belongs to a local vector space. However, the matter field itself is not local.

The description of derivatives of vector fields in gauge theories presents a problem. The reason is that derivatives are the local limit of the difference of field values at neighboring points. The indicated vector subtraction is not defined because the field values are in different vector spaces. Subtraction is defined only within a vector space structure. It is not defined between two different vector spaces.

This is remedied by the use of unitary maps between vector spaces at different points. As connections or elements of a unitary gauge group, these maps take vectors in a vector space at one point into vectors in a space at another point. Use of these maps enables derivatives of vector fields in gauge theories to be defined within a single local vector space. The Lie algebra generators of these unitary maps correspond to bosons. These bosons are the force carriers between fermions. The type and number of bosons depends on the dimensionality of the gauge group and on other things.

The success of gauge theories in physics is remarkable. They form the basis of the standard model. The many descriptions in the model of the electromagnetic, weak, and strong forces and their interactions with the elementary particles of the model are supported by experiment. However, as is well known, the standard model is incomplete. It does not include gravity, dark matter, dark energy, inflation, or other fields discussed in the literature. There are also a good number of arbitrary parameter values that must be used as input to the model.

The mathematics of gauge theories as used so far have a property that seems strange. This is based on the fact that scalars are a basic component of the axiomatic description of vector spaces.[1] However,

---

[1]For example, vector spaces are closed under multiplication of a vector by a scalar. Norms and scalar products of vectors have values in the associated scalar field.

local vector spaces are not accompanied by local scalars. Use of one global set of scalars (real or complex numbers) as the scalars for all the local vector spaces does seem strange. It seems appropriate to expand gauge theories to include local scalars. Exploration of the effects of this expansion, and the use of local scalars, and other local mathematical structures, in mathematics, physics, and geometry, is the topic of this book.

1.2   Summary of this book

One of the goals of this book is to expand the use of local mathematical systems to areas of physics and geometry other than that of vector spaces in gauge theories. This book is an expansion and extension to more areas of physics and geometry of work, begun in 2010 [18], continued in other work exemplified by [19] and [20] with [21] a recent work. The aim here is to explore the effect of such an expansion on theoretical descriptions of physical quantities. Hopefully, it might be of help to explain some of the outstanding problems in physics and the connections between physics and mathematics.

For this work, it is necessary to have some common description of mathematical systems of different types. This is especially useful for the description of mathematical systems localized to points in space or space–time. A useful common description is provided by the representation of different mathematical systems by structures and a set of axioms that are valid for the system type under consideration.

This representation of different types of mathematical systems as structures is based on the description in mathematical logic [22] of semantic models of syntactic systems. These models are structures with several different components. These consist of a base set, a few or no basic operations, a few or no basic relations, and a few or no constants. The structures, or models, are required to satisfy a set of axioms relevant to the system type under consideration.

This representation of mathematical systems as structures will be used throughout this book. In addition, the example provided by gauge theory with separate vector spaces at each space–time location will be expanded here to include separate mathematical structures of many different types at each space–time point. This results in the localization of mathematics.

By itself, the use of local mathematical structures instead of global ones does not affect theoretical descriptions of physical and geometric quantities. The effect arises from the discovery that two concepts, that of number and number value, are distinct. They are not conflated as is the usual practice in physics and mathematics. This is a consequence of the discovery that for each number type (natural numbers, integers, rationals, reals, and complex numbers) there are many different structures, each differing by a value or scaling factor. The result is that a rational number, such as the symbol string, 3.12, by itself has no intrinsic value. The value of the number 3.12 is determined by the scaling or value factor of the structure containing the number.

The mathematics of number structures of different types with different scaling factors is shown in some detail in the following chapter. Simple examples based on the natural numbers are given. There is also a discussion of the relationship between number structures of the same type but with different scaling factors. This will turn out to be important in the following chapters.

The effect of number scaling is not limited to just numbers. It also affects mathematical systems that include numbers of some type as part of their axiomatic descriptions. Mathematical systems, such as vector spaces, operator algebras, and group representations, are all affected by number scaling. Because of the importance of vector spaces to physics, the number scaling of these spaces is also described in the following chapter. Included is a description of the relation between vectors in spaces with different scaling factors.

The extension of local mathematics[2] with many structures of each number type with the structures differing by scaling or value factors is problematic for physics and geometry. The reason is that theory expressions and predictions, and descriptions of geometric systems are expressed in terms of meanings or values of numbers, vectors, and of other types of mathematical elements. However, theory computation outputs and experimental outcomes are numbers. They are not number values. Since the value of a number depends on the scaling or value factor of the structure containing the number, the

---

[2]The term, local mathematics, is used in a different way by [23]. This reference develops a category theoretic approach to set theory. The truth of some set theory axioms depends on the local framework for the sets.

value or meaning of these numbers is unknown. There is no connection between theory and experiment since the computer outputs and experiment outcomes can have any value. In addition, the relation between the values of numbers at different locations in space or time is not known.

This problem is fixed with the introduction of a space, time, or space–time-dependent scaling or value field. This field determines the value factor for the number and vector space structures at each location in a space, time, or space–time manifold, M.

In this book, the value field is restricted to be a real scalar field. The main reason that it is not a complex number valued field is that it is not clear how to handle geometric properties with a complex number value field. For gauge theories and quantum mechanics, there is no reason that the field cannot be complex [24].

The scalar field will be referred to as a scaling field, a value field, or a meaning field. These are different labels for the same field. Which name is used will depend partly on the context in which the field is used.

Support for the presence of this value field is provided by extension of the argument used by Yang and Mills [25] in gauge theories to support the existence of unitary gauge transformations between vector spaces at different locations. The original argument for isospin space, "which state represents a proton at one point does not determine which state represents a proton at another location" [25], is expanded here to "the value of a number at one location does not determine the value of the number at another location". This is an extension of the "No information at a distance" principle [16,17] to numbers of different types. The value field provides the base for a definition of connections as maps between mathematical structures at different locations. As used here, connections are the scalar equivalent of unitary gauge transformations for vector spaces.

The main effects of local mathematical structures and the presence of a value field show up in the theoretical descriptions of many physical properties and systems. Of special importance are those properties represented as integrals or derivatives of a field over space, time, or space–time. If one follows the prescription laid out for vector spaces in gauge theories, then the values of the field at each location belong to a local structure at the field location. Integration over the field is not defined because the addition of field values, implied in the definition of an integral, are in different structures. Addition is defined for elements within a structure, not between structures at different locations.

The same problem holds for derivatives. As seen in gauge theory, the definition of a derivative includes subtraction of field elements in local mathematical structures at neighboring locations. The subtraction is not defined because subtraction is defined for elements within a structure, not between elements in different structures.

Wave functions in quantum mechanics provide a simple example of these problems for integrals. A wave function is a field over space whose values are complex numbers. Addition or subtraction of wave function values at different locations is not defined because the values are numbers in different complex number structures at different locations.

The method used in gauge theory to solve the problem is extended here to scalar fields. The use of unitary gauge transformations as parallel transports of vectors between vector spaces is extended here to define scalar connections as parallel transports of numbers from one local number structure to another. These connections are number preserving and number value changing. As a result, they affect the theoretical description of some physical quantities.

The first two chapters of this book, as Part I, give details on the mathematical description of number structures of different types and vector spaces associated with different scaling or value factors.

The following chapter describes the effect of the distinction between number and number value for the different types of numbers. The description is detailed because the fact that numbers can have quite different values in different structures is hard to accept. This is the case even though the mathematics itself is not difficult. Relations between numbers and their values in structures with different value factors are described. These will be important in application of the mathematics of these structures to physics and geometry. The scaling is extended to vector spaces in that the length of vectors depends on the scaling factor for the number structure containing the vector length number.

Fiber bundles [27,28] provide a very suitable, almost intuitive mathematical arena for the representation of local mathematical structures at each location of a space or space–time manifold. There are no restrictions on the mathematical contents of the fibers. They can include as many different mathematical structures as are needed for the problem at hand.

Fiber bundles and connections are described in Chapter 2. The fibers at each manifold location contain local mathematical structures of different types and with different scale or value factors. Physical systems, described theoretically as extending over space or space–time, are represented with sections on the fiber bundle. Fermion fields in gauge theory, wave functions in quantum mechanics, and paths as tangent bundles are examples.

Parallel transports or connections play an important role in this book. They are needed to transport field components or path tangent bundle components to a common reference location. The connections are needed to define physical and geometric quantities as integrals or derivatives over space or space–time.

Part II describes the effect of a space or space–time varying value field on quantities in physics and geometry. The first chapter in this part gives the effect that the value field and local scalar structures have in gauge theory. The result is the introduction of a new vector field $\vec{A}$, as the gradient of a scalar value field, into the Dirac Lagrangian and other Lagrangians in gauge theories. A term for the interaction of $\vec{A}$ with the fermion fields is included. The value field is a scalar field presumably of spin 0. No mass restrictions on the value field are present. The observation that the value field shares these properties with the Higgs field is noted.

The following chapter describes the effect of local mathematics and the presence of the α field on some quantities in quantum mechanics. A description of the effects of the value field on position and momentum representations of one-particle states and scalar products is given. The effects of α show up in the use of parallel transport of these quantities, represented as sections on a fiber bundle, to a reference location. Scalar products of the localized states are shown to differ from localized scalar products by having an extra scaling factor in their description. Differences in the effect of the value field also show up in a comparison of expectation values of the position and momentum expectation values of localized states. Normalization, Hamiltonians, and quantum dynamics with the value field present are also discussed.

The description of one-particle states is followed by a discussion of two-particle states, including entangled states. Localization of two-particle entangled states requires an expansion in the use of local mathematics and the effects of the α field. The representation of an entangled state as a section on a fiber bundle requires the replacement of Hilbert spaces at each location by tensor products of pairs of Hilbert spaces at all pairs of space locations. The effect of the value field in parallel transport of the state components to a reference pair of Hilbert spaces is included by representing the effect by the arithmetic means of the field values at all pairs of locations [29]. A brief discussion of the extension of the effect of the value field to n-particle states concludes this chapter.

The effects of the value field and its gradient on properties of systems in Euclidean space, Euclidean space and time, and relativistic space–time are discussed in the chapter on geometry. The use of local mathematics requires that the mathematical structures in the fibers at each location in the base space or space–time include a local space or space–time.

Paths and their properties are described in the Euclidean geometries. In space–time, paths are timelike. Light is described in a section on lightlike paths. Two methods of lifting paths in space or space–time to image paths in local spaces or space–times are described. One method consists of lifts of whole paths to the local space or space–times. The other method represents paths in space or space–time by a bundle of path tangent vectors where the bundle is treated as a section on the fiber bundle. This corresponds to a lift of each vector in the bundle to a vector in the image space or space–time colocated with the tangent vector.

The emphasis in this work is on the second method of constructing local images of paths. The reason is that it follows the use of parallel transport in gauge theories to describe field derivatives. It emphasizes the observation that this book expands the constructions used in gauge theories to numbers and to large parts of mathematics and their effect on physics and geometry.

Determination of properties, such as lengths, of paths as sections requires parallel transport of the section tangent vectors to a local space or space–time at a reference location. The result is the formation of local representations of paths as local tangent vector bundles.

Distances between points in the local spaces of space–times are determined by finding the extremum for the lengths of paths between the points.

A feature of the effect of the value field in geometry is that changes in angles and trigonometric functions of angles change under parallel transport. But equations relating different functions are invariant under parallel transport. This invariance holds for equations in general, not only those related to geometric quantities.

Geodesic equations in space–time are affected by the gradient field, $\vec{A}$, of the value field. The deviation of these equations from those of a straight line shows that variations in the value field affect the geometries of the local space–times.

The deviation of the geometries of the local space–times from that of the base space, M, can also be expressed by metrics for the space–times. These are obtained from the usual one for flat space–time by adding an overall space–time location-dependent scaling factor. The effect of the differences in the geometries of the local space–times from that of the flat base space–time can be minimized by requiring that differences are too small to be observed in a local region of the universe. This is the region of the universe that is occupiable by us as observers.

The effect of variations in the value field on light shows up as a variation in the wavelength of light as it moves along a lightlike path. A simplification of the value field to depend on cosmological time only and not on space location is used.

Temporal variations in the value field can be such as to describe either red shifts or blue shifts of light arriving from distant light sources. An example of the value field is given that describes the cosmological red shift due to Hubble expansion of space. The example also shows that variations in the value field can be such as to give the accelerated red shift that has been ascribed to dark energy. The resulting time-dependent expansion of local space–times is also described.

Chapter 7 describes the steps in the process of comparison of theory predictions with experiment results. It is noted that theory comparison with experiment consist of comparison of numerical values of computer outputs, as theory predictions, with numerical values of experiment outcomes. Computer outputs and experiment outcomes, as physical systems in certain states, are interpreted as numbers.

The presence of the value field is essential for comparisons between numerical values of computer outputs with those of experiment outcomes. Without the value field the freedom of choice of values of numbers means that the values of the output and outcome numbers are not known. Theory experiment comparison is not possible.

With the value field present, the values of the numbers are known. Since computations and experiments are carried out in different space–time locations, comparison of the values of the output and outcome numbers requires parallel transport of the numbers to a common location.

As might be expected, local mathematics and the presence of the value field affect comparison of statistical theories with experiment. Repeated experiments yield outcome numbers at different locations in space or space–time. Determination of the mean of the values of the outcome numbers requires parallel transport of the individual outcome numbers for the calculation of the mean. Comparison of theory with experiment includes the additional complication arising from the location dependence of the values of the mean and the theory predictions.

The lack of direct experimental evidence for the presence of variations in the value field means that variations of the field in a local space–time region must be too small to be seen experimentally. The space–time region is the small part of the universe that is occupiable by us or others with whom communication is possible. There are no restrictions on the value field outside the region. The local conditions of the value field follow from the fact that experiments are local. They consist of state preparations of local systems followed by measurements.

Local measurements are not so restrictive. They can determine properties of systems far away at cosmological distances. For this reason, the locality condition does not conflict with the value field description of the Hubble expansion and possibly dark energy.

The last chapter emphasizes the essential role that the value field plays. This is the case even if the field is a constant everywhere. The reason is that physical representations of numbers, as symbol strings or as some other type of physic systems, have no intrinsic meaning. Any meaning is possible as the numbers can belong to any one of an infinite number of number structures of the appropriate type.

The presence of the value or meaning field gives value or meaning to any physical representation of a number. This is the case even if the value field is a constant everywhere. Its existence is implicit in that physical representations of numbers do have meaning or value.

The section of the last chapter is very speculative. It suggests that the meaning field may be essential for an observer to be conscious of or aware of the value of a number. It is based on the observation that there is a correspondence between different physical states of the brain and being conscious of the meaning of a number. If the different physical states of the brain have no intrinsic meaning, then the value field would be needed to give value or meaning to the physical brain states. This is equivalent to an observer being conscious or aware of the value or meaning of a number.

The overall framework or philosophical position taken in this work begins with the observation that physical systems that we can directly observe have no intrinsic meaning. They are physical systems in various states. We may observe them changing states as time passes, breaking up, joining together, or disappearing. We can describe what we see, hear, feel, and smell, but this is quite different from giving meaning to these observations.

Examples of these systems of concern here are numbers. They consist physically of string of number symbols or digits. We can tell from the form of the strings what type of number they are, integral, rational, real, or complex, but this does not yield the meaning or value of the symbol strings.

The function of the scaling or value field is to provide value or meaning to the numbers. These are used to create theoretical descriptions of physical systems and their dynamics and predictions of how the systems will evolve. Physical theory provides meaning to the otherwise meaningless behavior of physical systems in various states.

A useful analogy is to regard the meaning field as a decoding field. Numbers and, more generally, strings of alphabet symbols can represent codes for the values of numbers and for the meaning of strings of alphabet symbols. Decoding of numeral strings recovers the meaning or value of the numeral strings. Decoding of strings of alphabet symbols recovers their meaning.

The coding and decoding appropriate to this work are that encoding and decoding can depend on the space and/or time locations of the symbol strings. A string of numbers at one location may not have the same meaning at one location as at another location. The time and/or space rate of change in the coding and decoding is given by the time and/or space derivative of the meaning field.

In these structures, N, I, Ra, R, and C are base sets on which the basic operations $+, -, \times$, and $\div$ are defined with their properties given by the relevant axioms. The order relation is given by $<$ on all structures except $\bar{C}$. 0 and 1 are the elements of the base sets that satisfy the respective existing axioms for additive and multiplicative identities. Structures are distinguished from base sets by an overline, such as in $\bar{R}$ and R.

A representative structure for a normed vector space structure is given by

$$\bar{V} = \{V, +, -, \odot, |-|, \psi\}. \tag{1.1}$$

Here ψ denotes an arbitrary vector in the base set, V, of vectors, $|-|$ denotes a vector norm, and $\odot$ denotes scalar vector multiplication. Hilbert spaces are special examples of vector spaces in which the vector norm is replaced by a scalar product. The scalar vector product operation, norms, and scalar products are examples of maps between structures of different types, vector spaces, and scalar structures, such as $\bar{R}$ or $\bar{C}$. The structure, $\bar{V}$, is distinguished from the base set, V, by an overline.

1.3   The natural numbers

Natural numbers provide the simplest example of the distinction between number and number value. Consider the set

$$N = 0, 1, 2, 3, 4, \dots \tag{1.2}$$

The symbols "0", "1", "2", etc. are names of natural numbers. The numbers in the set, N, are also discretely well ordered. The position of a number in the well ordering determines the value of the number. The value of the number 0 is 0, the value of the number 1 is 1, and so on. The concepts of number and number value coincide.

The set N with the well ordering and the addition and multiplication operations is a representation of the natural numbers. The structure with N as a base set satisfies the axioms for the natural numbers.

Consider the subset

$$N_2 = 0, 2, 4, \ldots \qquad (1.3)$$

of even numbers. This set inherits the well ordering from N. The values of the numbers in N_2 are determined by their place in the well ordering. The value of the number 0 is 0, the value of 2 is 1, the value of the number 4 is 2, and so on.

The structure with N_2 as the base set, with addition and multiplication operations, is also a representation of the natural numbers. It also satisfies the natural number axioms. The multiplication in this structure is scaled relative to that in the structure with N as the base set. The addition operation is unchanged.

This is the simplest example that shows the distinction between number and number value. It shows that the value of a number depends on the environment containing it. In this case, the environment is the well ordering.

The distinction between number and number value can be expressed succinctly by

$$1_2 \equiv 2_1.$$

This equivalence says that the number 2 is represented as the number in the base set N_2 with value 1 in $\bar{N}^2$. It is also represented as the number in the base set N_1 with value 2 in $\bar{N}^1$. The number 2 belongs to the base sets, N_1 and N_2. The base set N_2 is a subset of the base set, N_1. Note that by itself the symbol 2 has no intrinsic meaning or value. Its value is determined by the number structure, $\bar{N}^1$ or $\bar{N}^2$, containing it.

This equivalence can be extended to other numbers by noting that for any n

$$(2n)_1 \equiv n_2.$$

The number with value n in the structure $\bar{N}^2$ is the same number as is the number with value $2n$ in the structure $\bar{N}^1$.

This description can be extended to apply to subsets of N_1 consisting of every nth element of N_1. In this case $n_1 \equiv 1_n$, the number with value n in the structure $\bar{N}^1$ has value 1 in the structure $\bar{N}^n$.

Here and from now on, number structures with N_n as a base set will be designated by $\bar{N}^n$. This nomenclature will be used for many different types of structures. The capital letter represents the structure type and the subscript the scaling factor.

The number of different values a number can have depends on the value of the number in $\bar{N}^1$. If the number has a prime number value, p, in this structure, then the only possible values of the number are p in this structure and 1 in $\bar{N}^p$.

The number of possible values a number can have depends on its factors. For example, the number 30 can have several different values. These are shown by the equivalences,

$$30_1 \equiv 15_2 \equiv 10_3 \equiv 6_5 \equiv 5_6 \equiv 3_{10} \equiv 2_{15} \equiv 1_{30}. \tag{1.4}$$

These equivalences need examination in more detail as they are an example of much that follows. The number 30 consists of two physical symbols, the digit 3 followed by the digit 0. As a string of symbols it has no intrinsic meaning or value. The shape and order of the symbols in the string show that the string is a number, but they say nothing about the meaning or value of the number.

The value is provided by the structure containing the number. For example, the number 30, represented by 15_2, has value 15 in $\bar{N}^2$. The number 30, represented by 10_3, has value 10 in $\bar{N}^3$. The number 30, represented by 30_1, has value 30 in $\bar{N}^1$.

The number 30 can also be regarded as a code for its meaning as a number value. There are several possible decodings, each determined by the scale factor of the number structure containing the number. The possible decodings are shown by the equivalences in Eq. (1.4).

The representation of number structures used in Eq. (1.4) can be extended to structures $\bar{N}^n$ for any n. The structure is given by

$$\bar{N}^n = \{N_n, +_n, \times_n, <_n, 0_n, 1_n\}. \tag{1.5}$$

The subscript, n on the two operations, the $<$ relation, and the two constants indicate that they are defined on the base set, N_n. Also the arithmetic axioms are valid for $\bar{N}^n$.

The values of numbers in N_n depend on their properties in the structure, $\bar{N}^n$. These values can be defined by a valuation function, v_n, that has N_n as a domain. If j is a number with representation a_n in $\bar{N}^n$, then

$$v_n(j) = v_n(a_n) = a.$$

If j also has a representation in N_m as the base set of $\bar{N}^m$, the value of j in $\bar{N}^m$ is given by $v_m(j)$. This value is different from $v_n(j)$.

It is important to note that the relations between numbers such as Eq. (1.4) are equivalences. Equality is not used. The reason is that the equal relation, the order relation, and arithmetic operations are not defined between different number structures. They are defined only within single number structures. This important point will be emphasized by repetition *ad nauseam* throughout this work.

The representation between numbers and their representations and values is closely related to the distinction between syntactic and semantic expressions in mathematical logic [22]. Number symbol strings and arithmetic combinations of numbers such as $27 + 45$ and 432×12 are meaningless strings. These are syntactic expressions. The set of all finite length strings of digits corresponds to the base set of numbers referred to above. The representation of a base set number, j, in a semantic model structure $\bar{N}^k$ is $[v_k(j)]_k$.[3] The base set, N_k, of $\bar{N}^k$ consists of the representations of all numbers j such that $v_k(j)$ is any one of the number values, $0, 1, 2, \ldots$. Here N_k is a subset of N_1.

The setup here is an expansion of that described in mathematical logic for standard models of natural numbers and arithmetic. The model, $\bar{N}^1$, for $k = 1$ corresponds to the usual standard model of the natural numbers. It is the only model containing a representation of all the number strings.

The expansion consists of the presence of additional scaled models, $\bar{N}^k$, one for each value of k. Each syntactic number, j, has semantic equivalences in all models $\bar{N}^k$, where k is a factor of the value, $v_1(j)$, of j in $\bar{N}^1$. Note that 0 is the only syntactic number that has a representation in the models for all values of k.

Syntactic arithmetic expressions such as $j + l$ and $j \times l$ have corresponding meaningful expressions in $\bar{N}^1$. They are given by

$$
\begin{aligned}
[v_1(j + l)]_1 &= [v_1(j)]_1 +_1 [v_1(l)]_1, \\
[v_1(j \times l)]_1 &= [v_1(j)]_1 \times_1 [v_1(l)]_1.
\end{aligned}
\tag{1.6}
$$

---

[3]The square brackets enclose the value of the number. The subscript gives the scale factor for the structure containing the numbers. This representation of numbers will be much used.

If j and l are relatively prime, then $\bar{N}^1$ is the only structure in which the valuation of arithmetic expressions makes sense. If j and l have k as a common factor, then the arithmetic expressions $j + l$ and $j \times l$ have meaningful expressions in $\bar{N}^k$. They are

$$
\begin{aligned}
[v_k(j + l)]_k &= [v_k(j)]_k +_k [v_k(l)]_k, \\
[v_k(j \times l)]_k &= [v_k(j)]_k \times_k [v_k(l)]_k.
\end{aligned}
\tag{1.7}
$$

Let m be a number value different from n. The structure $\bar{N}^m$ is defined by

$$
\bar{N}^m = \{N_m, +_m, \times_m, <_m, 0_m, 1_m\}.
\tag{1.8}
$$

The map v_m has the same properties for $\bar{N}^m$ as v_n has for $\bar{N}^n$.

Relations between numbers and their values in $\bar{N}^n$ and $\bar{N}^m$ are described by maps between the two structures. There are two types of maps of $\bar{N}^m$ to $\bar{N}^n$. One type is value preserving and number changing. The other type is number preserving and value changing.

Value preserving number changing maps are the usual type that come to mind. These maps are defined by

$$
+_m \to +_n, \quad \times_m \to \times_n, \quad <_m \to <_n, \quad 0_m \to 0_n, \quad 1_m \to 1_n
$$

or $\bar{N}^m \to \bar{N}^n$. The maps of numbers are defined by $a_m \to a_n$. For each number value a, the number a_n is different from a_m. There is one exception to this. 0_m is the same number as is 0_n.

The maps of interest here are number preserving value changing maps. For natural numbers, these maps are defined from $\bar{N}^m$ into $\bar{N}^n$ only if n is a factor of m. In this case, N_m is a subset of N_n. It follows that every element of N_m is an element of N_n.

From Eq. (1.4), one sees that the relation between the values of numbers in the structures, $\bar{N}^m$ and $\bar{N}^n$, is given by

$$
m v_m(j) = n v_n(j)
\tag{1.9}
$$

for all j in N_m. One also has

$$
v_m(a_m) = a = v_n(a_n).
\tag{1.10}
$$

Note that a_m is a different number than a_n. If $m \neq n$, then $a_m \neq a_n$.

These relations can be used to obtain the relation between the numbers a_m and a_n. From Eqs. (1.9) and (1.10), one has

$$v_n(a_n) = v_m(a_m) = \frac{n}{m} v_n(a_m) = v_n\left(\frac{n_m}{m_m} a_m\right) = v_n\left(\left[\frac{n}{m} a\right]_m\right).$$
(1.11)

Here n_m and m_m are the numbers in N_m and in N_n that have values, n and m in $\bar{N}^m$. This is expressed by

$$v_n(m_n) = m \text{ and } v_n(n_n) = n.$$
(1.12)

Equation (1.11) says that the number that has value a in $\bar{N}^n$ is the same as the number that has value $(n/m)a$ in $\bar{N}^m$. Since the value maps are isomorphisms, Eq. (1.11) gives[4]

$$a_n \equiv \frac{n_m}{m_m} a_m$$
(1.13)

as the relation between the numbers a_n and a_m. Equivalence is used instead of equality because the number on the left side of the equivalence is in a different number structure than is the number on the right. Figure 1.1 illustrates the relations between the value maps and numbers in structures for the factors m and n as shown in Eq. (1.11). The figure shows clearly that one cannot exchange m with n. The reason is that a_n is not in the set N_m of numbers for $\bar{N}^m$.

The relations between numbers and their values described so far are restricted to cases where n is a factor of m. This restriction is the basis of a problem. Suppose one wants to add to numbers a_n and b_m where both m and n, and a and b are relatively prime. An example is the sum $4_2 + 5_3$. This is the addition of the number with value 4 in $\bar{N}^2$ to the number 5_3 with value 5 in $\bar{N}^3$.

This addition is not defined for two reasons. One is that addition of numbers is defined only within number structures, not between

---

[4]The presence of the division operation may cause some concern as division is not a natural number operation. This should not be a problem because the discussion comparing numbers and their values in different number structures is not a discussion within any natural number structure.

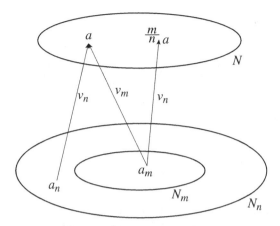

Figure 1.1. Representation of the relations between v_n and v_m on numbers in the base sets N, N_n, and N_m. Here N is the set of number values in $\bar{N}$ and $\bar{N}_m$ and N_n are the sets of numbers in $\bar{N}^m$ and $\bar{N}^n$. Also, a denotes a natural number value. Note that one cannot exchange the subscripts n and m as a_n is not in the domain of v_m.

number structures. This problem can be remedied by defining connections between number structures.

The other more fundamental problem is the fact that the number 4_2 is not a member of the base set, N_3, of numbers for the structure, $\bar{N}^3$. In this case, connections as number preserving value changing maps are of no help in removing this problem.

For natural numbers, this problem is resolved by noting that the number 1 is the only factor common to all natural numbers. Unrestricted addition of two numbers is valid for the structure $\bar{N}^1$ only. For this structure, concepts of number and number value are conflated.

The problem of disjoint base sets does not exist for number structures that are mathematical fields in that they are closed under division. For these types of numbers, rational real, and complex numbers, there is one base set. It is the same for all scaled representations for each number type.

The other problem, arithmetic operations are defined only within number structures, is remedied by the use of number preserving value changing connections. For rational, real, and complex numbers, the distinction between number and number value is relevant to mathematics, physics, and geometry.

1.4   The integers

The description for integers differs only in the presence of subtraction. The model corresponding to that of the natural numbers consists of a set I that is discretely linearly ordered with no minimal or maximal element. If j is a positive integer value, then the structure corresponding to $\bar{N}^n$ of Eq. (1.5) is

$$\bar{I}^j = \{I_j, +_j, -_j, \times_j, <_j, 0_j, 1_j\}. \tag{1.14}$$

If j is negative, then the order relation, $<_j$, is replaced by $>_j$.

The rest of the description of scaling for the integers follows that for the natural numbers. If j and k are two positive integer scaling factors and k is a factor of j, then the numbers in the base set, I_j, are contained in I_k. The only integer structure in which addition can be defined for all integer pairs is $\bar{I}^1$.

1.5   The rational numbers

Number structure scaling for the rational numbers differs from that for the natural numbers and integers in that the base set is the same for all the scaled structures. Let s be a positive rational number value. The components of a scaled rational number structure, Ra^s, are defined by

$$\overline{Ra}^s = \{B_{Ra^s}, \pm_s, \times_s, \div_s, <_s, 0_s, 1_s\}. \tag{1.15}$$

The reason for the superscript s on the base set will be explained soon.

The base sets, B_{Ra}, B_R, and B_C, for the rational, real, and complex numbers differ from that for the natural numbers and integers in that they are the same for all the rational, real, and complex number structures for all values of the scaling factor. This is a result of structures for these three number types being closed under division.

It follows from this that a rational number, by itself, has no intrinsic value. It can have any rational value. The value depends on the scaling factor for the structure containing it. This dependence of number value on the scaling factor can be expressed by a scale factor-dependent valuation function that assigns rational values to each number in B_{Ra}.

For each value or scale factor, s, the value map, v_s, is used to map each rational number in B_{Ra} to a number in the base set, B_{Ra^s}, of rational numbers for $\overline{Ra}^s$. If β is a rational number in B_{Ra}, then the corresponding number in B_{Ra^s} is $[v_s(\beta)]_s$. The value of this number is $v_s(\beta)$. The base set, B_{Ra^s}, is defined by

$$B_{Ra^s} = \cup_{\beta \epsilon B_{Ra}} [v_s(\beta)]_s. \tag{1.16}$$

The valuation function can be extended to relate arithmetic combinations of numbers in B_{Ra} to corresponding combinations of numbers in B_{Ra^s}. For each pair r, u of rational numbers,

$$[v_s(r \pm u)]_s = [v_s(r)]_s [v_s(\pm)]_s [v_s(u)]_s = [v_s(r)]_s \pm_s [v_s(u)]_s,$$
$$[v_s(r \times u)]_s = [v_s(r)]_s [v_s(\times)]_s [v_s(u)]_s = [v_s(r)]_s \times_s [v_s(u)]_s,$$
$$[v_s(r \div u)]_s = [v_s(r)]_s [v_s(\div)]_s [v_s(u)]_s = [v_s(r)]_s \div_s [v_s(u)]_s,$$
$$[v_s(r < u)]_s = [v_s(r)]_s [v_s(<)]_s [v_s(u)]_s = [v_s(r)]_s <_s [v_s(u)]_s. \tag{1.17}$$

The two numbers in the base set B_{Ra^s} that have values 0 and 1 in $\overline{Ra}^s$ are denoted by 0_s and 1_s.

The relation between the rational numbers in B_{Ra} and their arithmetic combinations and the rational numbers and their combinations in $\overline{Ra}^s$ is similar to the relation between syntactic and semantic structures in mathematical logic [22]. Rational numbers in B_{Ra} and their combinations correspond to syntactic expressions as meaningless symbol strings. Rational number structures correspond to semantic models of the syntactic expressions.

A brief summary of the structure of syntactic expression for numbers and other types of mathematical systems, as exemplified by the rational numbers, is useful. The numbers in B_{Ra} and their arithmetic combinations into terms are meaningless syntactic symbol strings. So are expressions that have open variables and are variable-free. (These are expressions in which all variables are arguments of "for all" and "there exists" as in "for all x ..." and "there exists a y ...".) Besides, symbols for "for all" and "there exists" are symbols for "or", "and", "not", and = or equal. There is also a set of expressions that are axioms for the type of number under consideration. Proofs are strings of expressions that are obtained from the axioms by use of the rules for logical deduction. Any expression in a proof is a theorem.

Syntactic expressions correspond to meaningful semantic expressions. These are true or false. All syntactic theorems correspond to true expressions in the model.

In the setup used in this work, there are many semantic or meaningful rational number models corresponding to a syntactic structure for the numbers. For each value of s, $\overline{Ra}^s$ is a model. This contrasts with the usual relation in which there is one standard semantic model associated with the syntactic structure. This is the model for $s = 1$.

Let t be a positive rational number value with

$$\overline{Ra}^t = \{B_{Ra^t}, \pm_t, \times_t, \div_t, <_t, 0_t, 1_t\} \tag{1.18}$$

the associated scaled structure. The corresponding isomorphism v_t from $\overline{Ra}^t$ to $\overline{Ra}$ has the components shown in

$$v_t(B_{Ra}) = \overline{Ra}, \ v_t(\pm_t) = \pm, \ v_t(\times_t) = \times$$
$$v_t(\div_t) = \div, \ v_t(<_t) = <, \ v_t(0_t) = 0, \ v_t(1_t) = 1. \tag{1.19}$$

The base set B_{Ra^t} is defined by Eq. (1.16) with t replacing s.

Relations between the isomorphisms, $v_s : \overline{Ra}^s \rightarrow \overline{Ra}$ and $v_t : \overline{Ra}^t \rightarrow \overline{Ra}$, are given by

$$v_s(a_s) = v_t(a_t) = a \tag{1.20}$$

and

$$v_s(a_t) = \frac{t}{s} v_t(a_t) = v_t\left(\frac{t_t}{s_t} a_t\right) = \frac{t}{s} a. \tag{1.21}$$

Here s_t and t_t are rational numbers in $\overline{Ra}^t$ that have values s and t in $\overline{Ra}$. The relation obtained by exchanging s and t is

$$v_t(a_s) = \frac{s}{t} v_s(a_s) = v_s\left(\frac{s_s}{t_s} a_s\right) = \frac{s}{t} a. \tag{1.22}$$

These results also show that if $s = -t$, then $a_s = -a_t$.

This representation of rational numbers in the form of number values with the scale factor as a subscript is very useful. It will be much used in this work. However, it is redundant. It is obvious that $v_s(a_s) = a$. This expression lacks the connection between numbers in the base set, B_{Ra}, and number expressed as a_s. The connection is made explicit by letting β be a rational number in B_{Ra} such that $v_s(\beta) = a$. Then $v_s(a_s) = a$ is equivalent to

$$v_s([v_s(\beta)]_s) = v_s(\beta). \tag{1.23}$$

Note that arithmetic operations are defined in $\overline{Ra}^s$ on elements of the form $[v_s(\beta)]_s$. They are not defined on elements of B_{Ra}.

As a specific example, the rational number, 3.42, in B_{Ra} has value $v_r(3.42)$ in $\overline{Ra}^r$. The number in $\overline{Ra}^r$ with this value is $[v_r(3,42)]_r$.

One can proceed with this representation of rational numbers with their values represented by a valuation function without ever mentioning what the specific value of a rational number in a rational number structure is. The value of 3.42 in the rational number structure $\overline{Ra}^r$ is $v_r(3.42)$. This representation holds for all scale factors, r. This representation says nothing about what specific value is represented by $v_r(3.42)$.

The distinction between numbers and number values for the natural numbers shows a solution to this problem. For natural numbers, it was seen that for any number n in the base set, the value of n in $\bar{N}^1$ was n. Equation (1.4) gives an example of this. For $\bar{N}^1$, the concepts of number and number value are conflated.

It seems reasonable to extend this observation for natural numbers to apply to all number types. This can be expressed in the representation of numbers used here, such as a_s, by the equation

$$a_1 \equiv a. \tag{1.24}$$

This equation says that the rational number a_1 is identified with its value, a, in $\overline{Ra}^1$. Number and number value are conflated in the structure. An equivalent statement is that

$$\beta \equiv [v_1(\beta)]_1 \equiv v_1(\beta). \tag{1.25}$$

The number β can be identified with its value in $\overline{Ra}^1$.

The properties of the valuation function can be used to define the value of the number a in structures with any nonzero positive rational scaling factor. From Eq. (1.21), one has the relation

$$v_s(a_1) = \frac{1}{s}a = \frac{a}{s}. \tag{1.26}$$

Setting $\beta = a_1$ gives

$$v_s(\beta) = \frac{1}{s}\beta = \frac{\beta}{s}. \tag{1.27}$$

Equation (1.25) is used to obtain this result.

As a specific example, the value of the rational number 3.42 is 3.42 in $\overline{Ra}^1$. The value of this number in the structure $\overline{Ra}^{6.43}$ is 3.42/6.43. If a rational number b in the base set has the value of b in $\overline{Ra}^1$, then the value of b in $\overline{Ra}^r$ is b/r. Here r is a rational scaling or value factor.

The number 0 is special in that it is the only number that can be conflated with its value in all rational number structures. This can be expressed by the relation

$$0_s = 0 \tag{1.28}$$

for all scaling factors, s. This equation is also valid for real and complex numbers.

Addition between numbers in different rational number structures is not defined. However, because the base set B_{Ra} is common to all rational number structures, connections between structures fix the problem. As an example suppose a_s and b_t are numbers in $\overline{Ra}^s$ and $\overline{Ra}^t$. The relations between value functions for different scaling factors show that a_s is equivalent to the number $[sa/t]_t$ in $\overline{Ra}$. Addition to b_t is defined, giving the result $[sa/t + b]_t$.

Figure 1.2 is a summary of the relations between the maps and rational numbers shown in Eqs. (1.20)–(1.22). The figure shows the number f in the base set as the numbers a_s and b_t in $\overline{Ra}^s$ and $\overline{Ra}^t$. The number a_s is equivalent to b_t; they are the same numbers in their respective number structures. The two number value equations, $b = (t/s)a$ and $a = (s/t)b$, follow from the observations that number preserving value changing transformations of a_s and b_t give $a_s \rightarrow [\frac{s}{t}a]_t = b_t$ and $b_t \rightarrow [\frac{t}{s}b]_s = a_s$.

As was the case for the natural numbers, there are two types of maps that take $\overline{Ra}^t$ to $\overline{Ra}^s$. One is number changing and value

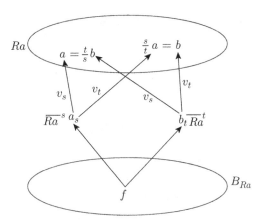

Figure 1.2. A summary of the relations between rational numbers and their representations and relations in rational number structures with different scaling factors. The sets B_{Ra} and Ra are the base set of rational numbers and the set of all rational number values. a and b denote the values of f in the rational number structures with scaling factors s and t.

preserving. The other is number preserving and value changing. Both maps preserve the validity of the axioms for rational numbers. The closure of rational numbers under inversion has the consequence that the maps of each type are elements of a group.

The number changing value preserving group of maps is represented by W_{Ra}. For each positive rational number value d, the group element $W_{Ra}(d)$ maps $\overline{Ra}^t$ isomorphically onto $\overline{Ra}^{dt}$. The map is defined by

$$W_{Ra}(d)\overline{Ra}^t = \overline{Ra}^s, \tag{1.29}$$

where $s = dt$ and

$$W_{Ra}(d)(a_t) = a_s, \ W_{Ra}(d)(\pm_t) = \pm_s, \ W_{Ra}(d)(\times_t) = \times_s,$$
$$W_{Ra}(d)(\div_t) = \div_s, \ W_{Ra}(d)(<_t) = <_s, \ W_{Ra}(d)(0_t) = 0_s, \tag{1.30}$$
$$W_{Ra}(d)(1_t) = 1_s.$$

Because $W_{Ra}(d)$ preserves values and operations, it makes $\overline{Ra}^t$ and $\overline{Ra}^s$ appear the same to all practical purposes. Also because it preserves number values, it has no effect on the values of physical quantities. This is probably the reason that number structure scaling has not been noticed in the literature. However, a_s and a_t are different numbers.

The group W_{Ra} is commutative and transitive with $W_{Ra}(d^{-1})$ the inverse of $W_{Ra}(d)$. These properties are a straightforward consequence of the definition of Eq. (1.29). Let d and e be two rational scaling values. Then,

$$W_{Ra}(d)W_{Ra}(e) = W_{Ra}(de) = W_{Ra}(ed). \tag{1.31}$$

The number preserving and value changing maps also form a group, Z_{Ra}, with elements $Z_{Ra}(d)$ for each rational number value $d = s/t$. These maps are defined by

$$Z_{Ra}(d)\overline{Ra}^t = \overline{Ra}^t_{dt}. \tag{1.32}$$

These maps can be described as number preserving environment changing maps. They map numbers and operations in the environment provided by $\overline{Ra}^t$ to the environment provided by $\overline{Ra}^s$. This change in environment gives a corresponding change in the values of numbers and operations.

The effect of environment change is seen from the action of $Z_{Ra}(d)$ on the components of $\overline{Ra}^t$. The action is given by

$$Z_{Ra}(d)(a_t) = \frac{t_s}{s_s}a_s \text{ for all rational number values, } a,$$

$$Z_{Ra}(d)(\pm_t) = \pm_s, \ Z_{Ra}(d)(\times_t) = \frac{s_s}{t_s}\times_s, \ Z_{Ra}(d)(\div_t) = \frac{t_s}{s_s}\div_s,$$

$$Z_{Ra}(d)\left(\frac{<_t}{>_t}\right) = \frac{<_s}{>_s}, \ Z_{Ra}(d)(0_t) = 0_s, \tag{1.33}$$

$$Z_{Ra}(d)(1_t) = \frac{t_s}{s_s}1_s.$$

As before, to save on notation, $s = dt$. If d is negative, $<_s$ and $>_s$ are interchanged in the definition.

The definition of $Z_{Ra}(d)$ can be summarized in the components of $\overline{Ra}^t_s$. One has

$$\overline{Ra}^t_s = \left\{ B_{Ra}, \pm_s, \frac{s_s}{t_s}\times_s, \frac{t_s}{s_s}\div_s, <_s, 0, \frac{t_s}{s_s}1_s \right\}$$

$$= \left\{ B_{Ra}, \pm_s, \left(\frac{s}{t}\times\right)_s, \left(\frac{t}{s}\div\right)_s, <_s, 0, \left(\frac{t}{s}1\right)_s \right\}. \tag{1.34}$$

This equation gives two equivalent representations of $\overline{Ra}_s^t$. The first equation gives the representation of the components of $\overline{Ra}^t$ in terms of the components of $\overline{Ra}^s$. The second equation moves the scaling factors for each component inside the valuation parentheses. Both equations satisfy the rational number axioms. Equation (1.28) is used to replace 0_s by 0.

In these equations, the direction of the order relation depends on the sign of t/s. If t/s is negative, the direction of the order relation, $<$, is reversed.

One property of the scale change map, $Z_{Ra}(d)$, is that it commutes with addition and subtraction. However, it does not commute with multiplication or division. Let a_t and b_t be two numbers in $\overline{Ra}^t$ with values a and b. For addition and subtraction, one has

$$Z_{Ra}(d)(a_t \pm_t b_t) = Z_{Ra}(d)(a_t)Z_{Ra}(d)(\pm_t)Z_{Ra}(d)(b_t)$$

$$= \frac{t_s}{s_s}a_s \pm_s \frac{t_s}{s_s}b_s \qquad (1.35)$$

$$= Z_{Ra}(d)(a_t) \pm_s Z_{Ra}(d)(b_s).$$

Here, as before, $s = dt$.

This equation shows that the action of the number preserving value changing map on the sum, $a_t \pm_t b_t$, gives the same result as does first implementing the map on a_t and b_t and then adding the resulting numbers. This shows that $Z_{Ra}(d)$ commutes with addition and subtraction.

For multiplication, one has

$$Z_{Ra}(d)(a_t) \times_t b_t) = Z_{Ra}(d)(a_t)Z_{Ra}(d)(\times_t)Z_{Ra}(d)(b_t)$$

$$= \frac{t_s}{s_s}a_s \frac{s_s}{t_s} \times_s \frac{t_s}{s_s}b_s = \frac{t_s}{s_s}(a_s \times_s b_s). \qquad (1.36)$$

However,

$$Z_{Ra}(d)(a_t) \times_s Z_{Ra}(d)(b_t) = \left(\frac{t_s}{s_s}\right)^2 (a_s \times_s b_s). \qquad (1.37)$$

Comparison of the two multiplication outcomes shows that the number preserving value changing map does not commute with multiplication.

Replacement of multiplication with division in the above two equations shows a similar lack of commutativity:

$$Z_{Ra}(d)(a_t \div_t b_t) = \frac{t_s}{s_s}(a_s \div_s b_s) \tag{1.38}$$

and

$$Z_{Ra}(d)(a_t) \div_s Z_{Ra}(d)(b_t) = a_s \div_s b_s. \tag{1.39}$$

These results show that which order of operations one uses, multiplication or division before scale change or these operations after scale change, depends on the context of the problem being considered. Both orderings are equally valid. It will turn out that determining the order of operations is not a problem. The ordering will be dictated by the problem being considered. Note that the ordering commutes if $s = t$.

The ordering, multiplication or division, then scaling, has the property that polynomial equations are preserved under scale change. This can be seen by first considering the term, $(a_t)^m \div_t (b_t)^n$. The action of $Z_{Ra}(d)$ on this term gives

$$Z_{Ra}(d)\big((a_t)^m \div_t (b_t)^n\big)$$

$$= Z_{Ra}(d)(a_t^m)Z_{Ra}(d)(\div_t)Z_{Ra}(d)(b_t^n) = \frac{t_0}{s_s}(a_s^m \div_s b_s^n). \tag{1.40}$$

This follows from the cancellation of scale ratios in the numerator and denominator. For the numerator, the $m - 1$ factors, s_s/t_s, from scaling the multiplication operations combine with m factors, t_s/s_s, from the scaling of the m a_t. The single remaining scale factor, t_s/s_s, is canceled by the same factor appearing in the denominator of the fraction. The remaining factor, t_s/s_s, comes from the scaling of the division operation.

This result, used inductively, shows that all terms as combinations of numbers, variables, and the basic operations, no matter how complicated, are all multiplied by a single factor, t_s/s_s, when converted from terms in $\overline{Ra}^t$ to terms in $\overline{Ra}^s$. It follows that polynomial equations are preserved under the action of the scale change operator, $Z_{Ra}(d)$.

Let $P_t = 0$ be a polynomial equation with terms in the polynomial made up of numbers, variables and operations, all components of $\overline{Ra}^t$.

Acting on both sides of the equation with the scale change operator $Z_{Ra}(d)$ gives, with $s = dt$,

$$Z_{Ra}(d)(P_t) = Z_{Ra}(d)(0) \leftrightarrow \frac{t_s}{s_s} P_s = \frac{t_s}{s_s} 0 \leftrightarrow P_s = 0. \tag{1.41}$$

This shows that polynomial equations are preserved under scale change.

1.6 The real numbers

The description of number structure scaling for the real numbers is very much like that for the rational numbers. Let s and t be two real number values. The real number structures scaled by s and t are given by

$$\begin{aligned}
\bar{R}^s &= \{B_{R^s}, \pm_s, \times_s, \div_s, <_s, 0_s, 1_s\}, \\
\bar{R}^t &= \{B_{R^t}, \pm_t, \times_t, \div_t, <_t, 0_t, 1_t\}.
\end{aligned} \tag{1.42}$$

The description of real numbers in B_{R^s}, B_{R^t} and the value maps are similar to those for the rational numbers. As examples, a_s and a_t are real numbers in $\bar{R}^s$ and $\bar{R}^t$ that have the value a. As expected, a_s and a_t are different real numbers. Also

$$B_{R^s} = \cup_{\beta \epsilon B_R} [v_s(\beta)]_s. \tag{1.43}$$

B_{R^t} is defined by replacing s with t in this equation.

The value maps, v_s and v_t, are isomorphisms, $v_s : \bar{R}^s \to \bar{R}$ and $v_t : \bar{R}^t \to \bar{R}$. The definitions are essentially the same as the definitions for rational numbers, in Eqs. (1.20)–(1.22). One has

$$v_s(a_s) = v_t(a_t) = a \tag{1.44}$$

and

$$v_s(a_t) = \frac{t}{s} v_t(a_t) = v_t \left(\frac{t_t}{s_t} a_t \right) = \frac{t}{s} a. \tag{1.45}$$

Here s_t and t_t are real numbers in $\bar{R}^t$ that have values s and t. From Eq. (1.44), one also has

$$v_s(a_t) = \frac{t}{s} v_s(a_s) = v_s \left(\frac{t_s}{s_s} a_s \right) = \frac{t}{s} a. \tag{1.46}$$

This gives

$$a_t \equiv \frac{t_s}{s_s} a_s. \qquad (1.47)$$

Exchanging s and t in Eq. (1.46) gives

$$v_t(a_s) = \frac{s}{t} v_t(a_t) = v_t \left(\frac{s}{t} t_t a_t \right) = \frac{s}{t} a. \qquad (1.48)$$

These results also show that if $s = -t$, then $a_s \equiv [-a]_t$.

The representation of rational numbers, in terms of their values in different structures, can be applied to real numbers. If f is a real number in the base set, B_R, and $v_r(f)$ and $v_s(f)$ are the values of f in $\bar{R}^r$ and $\bar{R}^s$, then $[v_r(f)]_r$ is a number in $\bar{R}^r$ that represents f and $[v_s(f)]_s$ is a number in $\bar{R}^s$ that represents f. One also has the equivalence

$$[v_r(f)]_r \equiv \left[\frac{r}{s} v_r(f) \right]_s = [v_s(f)]_s. \qquad (1.49)$$

In this equation, $v_r(f)$ is different from $v_s(f)$. Figure 1.2 for rational numbers also holds for real numbers.

The use of the valuation function to determine the relation between numbers in B_R is quite general. However, assigning specific values to numbers requires that a specific value function be chosen. The same definition used for the rational and natural numbers will be applied here. This is represented by the equation

$$v_1(f) = f \text{ or } f_1 = f. \qquad (1.50)$$

This says that for any real number f the value of f is f in $\bar{R}^1$. Number and number value are conflated in $\bar{R}^1$.

From this, one learns from Eq. (1.45) that

$$v_s(f_1) = \frac{f}{s}. \qquad (1.51)$$

The representation of the number f in $\bar{R}^s$ is $[(f/s)]_s$.

As was the case for the rational numbers, there are two types of isomorphic maps between scaled structures. One type changes numbers but preserves number values. The other type preserves numbers but changes values. Each of these map types forms a commutative, transitive group. For each scale changing real number value d, the number changing group, W_R, has elements, $W_R(d)$. The value changing group, Z_R, has elements $Z_R(d)$. As was the case for the rational numbers, $Z_R(d)$ is an environment changing map of numbers and operations.

The definitions of these maps are similar to those in Eqs. (1.29–1.33). For $W_R(d)$, one has

$$W_R(d)\bar{R}^t = \bar{R}^s. \tag{1.52}$$

Here and from now on in much of the section on real numbers $s = dt$. This is done to save on notation. Details of the map $W_R(d)$ are

$$W_R(d)(a_t) = a_s, \quad W_R(d)(\pm_t) = \pm_s, \quad W_R(d)(\times_t) = \times_s,$$
$$W_R(d)(\div_t) = \div_s, \quad W_R(d)(<_t) = <_s, \tag{1.53}$$
$$W_R(d)(0_t) = 0_s, \quad W_R(d)(1_t) = 1_s.$$

The change in numbers is seen from

$$a_s \equiv \frac{s_t}{t_t} a_t. \tag{1.54}$$

This equation shows that a_s is a different number from a_t. Value preservation follows from

$$v_s(a_s) = v_t(a_t) = a. \tag{1.55}$$

The properties of $Z_R(d)$ are the same as those for the rational numbers. Equation (1.33) is used to repeat the properties here for the real numbers:

$$Z_R(d)(a_t) = \frac{t_s}{s_s} a_s \text{ for all values, } a \text{ in } \bar{R}, \quad Z_R(d)(\pm_t) = \pm_s,$$

$$Z_R(d)(\times_t) = \frac{s_s}{t_s} \times_s, \quad Z_R(d)(\div_t) = \frac{t_s}{s_s} \div_s, \quad Z_R(d)\begin{pmatrix} <_t \\ >_t \end{pmatrix} = \begin{matrix} <_s \\ >_s, \end{matrix}$$

$$Y Z_R(d)(0_t) = 0_t = 0_s, \quad Z_R(d)(1_t) = \frac{t_s}{s_s} 1_s.$$

$$\tag{1.56}$$

If d is negative, $<_s$ and $>_s$ are interchanged in the definition. Also if $d = -1$, then

$$a_{-t} = -1_t(a_t) = (-a)_t. \tag{1.57}$$

The components of the relativized number structure, $\bar{R}_s^t$, defined by

$$Z_R(d)\bar{R}^t = \bar{R}_s^t \tag{1.58}$$

are shown in

$$\bar{R}_s^t = \left\{ B_R, \pm_s, \frac{s_s}{t_s} \times_s, \frac{t_s}{s_s} \div_s, <_s, 0_s, \frac{t_s}{s_s} 1_s \right\}. \tag{1.59}$$

As was the case for the rational numbers, the structures, $\bar{R}^t$, $\bar{R}^s$, and $\bar{R}_s^t$, are required to satisfy the axioms for real numbers.

The direction of the order relation in the three structures, $\bar{R}^t$, $\bar{R}^s$, and $\bar{R}_s^t$, depends on the value of d. If d is positive, then the direction of the order relation in $\bar{R}_s^t$ is the same as that in $\bar{R}^t$. If d is negative, then the direction of the order relation in $\bar{R}^t$ is opposite that in $\bar{R}^t$.

As was the case for rational numbers, the action of $Z_R(d)$ on terms of the general form

$$\frac{\prod_{j=1}^{n}(a_j)_t^{n_j}}{\prod_{k=1}^{m}(b_k)_t^{m_j}} \tag{1.60}$$

is given by

$$Z_R(d)\frac{\prod_{j=1}^{n}(a_j)_t^{n_j}}{\prod_{k=1}^{m}(b_k)_t^{m_k}} = \frac{t_s}{s_s}\frac{\prod_{j=1}^{n}(a_j)_s^{n_j}}{\prod_{k=1}^{m}(b_k)_s^{m_k}}. \tag{1.61}$$

The scaling factors of the multiplication operations cancel all but one of the scaling factors that come from the relations $(a_j)_t \equiv (t_s/s_s)(a_j)_s$ and $(b_k)_t \equiv (t_s/s_s)(b_k)_s$. These scaling factors cancel in the division operation. The single remaining scaling factor comes from the scaling of the division operation.

So far, the description of number structure scaling for the real numbers has followed that of the rational numbers. However, unlike the rational numbers, the real numbers are closed under the limit process. Axiomatically, they are complete. This adds some new properties. For example, if $f_t : \bar{R}^t \to \bar{R}^t$ is an analytic function, then

$$Z_R(d)f_t(a_t) = \frac{t_s}{s_s}f_s(a_s). \tag{1.62}$$

This equation shows the effect of environment change, from $\bar{R}^t$ to $\bar{R}^s$ on the number, $f_t(a_t)$.

Equation (1.62) follows from the fact that analytic functions can be represented as convergent power series. Equation (1.61) shows that the mapping of a power series term in $\bar{R}^t$ to a term in $\bar{R}^t_s$ multiplies the term by the same factor, t_s/s_s. This factor can then be brought outside as a common multiplier of the series to give Eq. (1.62).

Functions, like f, provide good examples of the fact that, in the presence of scaling, environmental changes for numbers do not commute with basic operations of multiplication and division. Equation (1.62) shows the effect of change of environment from $\bar{R}^t$ to $\bar{R}^s$ on the number, $f_t(a_t)$. A different number is obtained if one proceeds in a different order. First changing the environment on the argument of the function and then determining the value of the function give the number, $f_s((t_s/s_s)a_s)$, in $\bar{R}^s$. This is a different number than is $(t_s/s_s)f_s(a_s)$ for most functions.

The difference between the origin of these two numbers can be expressed by the use of $Z_R(d)$, valuation, and inverse valuation maps by

$$f_t(a_t) \to Z_R(d)(f_t(a_t)) = \frac{t_s}{s_s}(v_s^{-1}v_tf_t)(v_s^{-1}v_t(a_t)) = \frac{t_s}{s_s}f_s(a_s) \tag{1.63}$$

and

$$f_t(a_t) \to (v_s^{-1}v_tf_t)(Z_R(d)(a_t)) = f_s\left(\frac{t_s}{s_s}a_s\right). \tag{1.64}$$

The difference between the numbers in these two equations shows that there is a difference between the results of how one maps functions and their values as numbers from $\bar{R}^t$ to $\bar{R}^s$. One cannot say one method is right and the other is wrong. Instead, the method chosen must depend on the particular problem under consideration. Many examples of this will be seen in the chapters of Part II.

As was the case for rational numbers, equations are preserved under the action of $Z_R(d)$. This is shown by

$$f_t(a_t) = b_t \to Z_R(d)(f_t(a_t)) = Z_R(d)(b_t)$$

$$\to \frac{t_s}{s_s} f_s(a_s) = \frac{t_s}{s_s}(b_s) \to f_s(a_s) = b_s. \tag{1.65}$$

An equivalent property is the fact that equations are preserved under value map changes. This follows from the relations

$$v_t(f_t(a_t)) = v_t(b_t) \leftrightarrow f(a) = b$$

$$\leftrightarrow \frac{t}{s} v_s(f_s(a_s)) = \frac{t}{s} v_s(b_s) \leftrightarrow v_s(f_t(a_t)) = v_s(b_t). \tag{1.66}$$

Integrals and derivatives also scale by the same factor as do the analytic functions. If $f_t : \bar{R}^t \to \bar{R}^t$ is an integrable function, then

$$Z_R(d) \left(\int f_t(a_t) da_t \right) = \frac{t_s}{s_s} \int f_s(a_s) da_s. \tag{1.67}$$

The proof of this follows from

$$Z_R(d) \left(\int f_t(a_t) da_t \right) = \int Z_R(d)(f_{a_t}(a_t) da_t)$$

$$- \int (Z_R(d)(f_t(a_t)) Z_R(d)(\times_t)$$

$$\times Z_R(d)(da_t) \tag{1.68}$$

$$= \frac{s_s}{t_s} \int f_s(a_s) \times_s da_s.$$

One has a similar result for derivatives. If f_t is a differentiable function, then

$$Z_R(d) \left(\frac{df_t(a_t)}{da_t} \right) = Z_R(d) \left(\frac{f_t(a_t + da_t) -_t f_t(a_t)}{da_t} \right)$$

$$= [Z_R(d)(f_t(a_t + da_t)) -_s Z_R(d)(f_t(a_t))]$$

$$\times Z_R(d)(\div_t) Z_R(d)(da_t) = \frac{t_s}{s_s} \frac{df_s(a_s)}{da_s}. \tag{1.69}$$

Here $\lim da_t \to 0$ is implied.

The proof that W_R and Z_R are transitive, commutative groups is given in the Appendix.

1.7 Complex numbers

The description of number structure scaling for complex numbers is somewhat more involved than that for the rational and real numbers. This is a consequence of the fact that the scaling factors can be complex.

Let s and t be two complex number values. The components of the scaled complex number structures, $\bar{C}^t$ and $\bar{C}^s$, are given by

$$\bar{C}^t = \{B_{C^t}, \pm_t, \times_t, \div_t, {}^{*t}, 0_t, 1_t\},$$
$$\bar{C}^s = \{B_{C^s}, \pm_s, \times_s, \div_s, {}^{*s}, 0_s, 1_s\}. \tag{1.70}$$

The base sets B_{C^t} and B_{C^s} of complex numbers are defined from B_C by

$$B_{C^s} = \cup_{\beta \epsilon B_C}[v_s(\beta)]_s = [\bar{C}]_s \tag{1.71}$$

and

$$B_{C^t} = \cup_{\beta \epsilon B_C}[v_t(\beta)]_t = [\bar{C}]_t. \tag{1.72}$$

Here $\bar{C}$ represents the set of all complex number values.

The axiom set for complex numbers [34] does not require complex conjugation as a basic operation. It is included here in order to show the effects of scaling on this operation.

The equations representing the properties of the isomorphic value maps are much the same as those for the real numbers as in Eqs. (1.44)–(1.48). They are repeated here for convenience:

$$v_s(a_s) = v_t(a_t) = a \tag{1.73}$$

and

$$v_s(a_t) = \frac{t}{s} v_t(a_t) = v_t\left(\frac{t_t}{s_t} a_t\right) = \frac{t}{s} a. \tag{1.74}$$

Here s_t and t_t are complex numbers in $\bar{C}^t$ that have values s and t in $\bar{C}$. The relation obtained by exchanging s and t in Eq. (1.74) is

$$v_t(a_s) = \frac{s}{t} v_s(a_s) = v_s\left(\frac{s_s}{t_s} a_s\right) = \frac{s}{t} a. \tag{1.75}$$

Use of Eq. (1.73) to replace $v_t(a_t)$ with $v_s(a_s)$ in Eq. (1.74) gives

$$a_t \equiv \frac{t_s}{s_s} a_s. \tag{1.76}$$

The opposite replacement in Eq. (1.75) gives

$$a_s \equiv \frac{s_t}{t_t} a_t. \qquad (1.77)$$

These equations show that the right-hand and left-hand terms are different representations of the same number in B_C.

These results also show that if $s = t^*$, then

$$a_{t^*} = \frac{(t^*)_t}{t_t} a_t. \qquad (1.78)$$

This is a different number from $(a^*)_t$. This follows from $v_t(a_{t^*}) = (t^*/t)a$, but $v_t((a^*)_t) = a^*$. Equation (1.78) says that the complex number a_{t^*} that has value a in $\bar{C}^{t^*}$ has value $(t^*/t)a$ in $\bar{C}^t$.

As was noted before, one can proceed without making a specific choice for the value function. The description of number and number value for the natural numbers shows that number and number value are conflated for the scaling factor equal to 1. Use of this property for complex numbers means that for every complex number β in B_C, $v_1(\beta) \equiv \beta$. If $v_1(\beta) = a_1$ in $\bar{C}^1$, then conflation of number and number value is expressed by

$$a_1 \equiv a. \qquad (1.79)$$

Equation (1.76) with $t = 1$ gives

$$a \equiv a_1 \equiv \frac{t_s}{s_s} a_s = \left[\frac{a}{s} \right]_s. \qquad (1.80)$$

Also for all complex scaling factors, s, $0_s = 0$.

As was the case for the rational and real numbers, there are number changing, value preserving, and number preserving, value changing maps. For each complex number value d, $W_C(d)$ is a number changing value preserving map. It is defined by

$$W_C(d)\bar{C}^t = \bar{C}^{dt}, \qquad (1.81)$$

where

$$W_C(d)(a_t) = a_s, \; W_C(d)(\pm_t) = \pm_s, \; W_C(d)(\times_t) = \times_s,$$
$$W_C(d)(\div_t) = \div_s, \; W_C(d)(*_t) = *_s, \qquad (1.82)$$
$$W_C(d)(0_t) = 0_s, \; W_C(d)(1_t) = 1_s.$$

Note that a_t is different from a_s. Here, and in much of the rest of this section, $s = dt$.

The definition of number preserving value changing maps is similar to that for the real numbers. For each complex number value, d, the action of the map $Z_C(d)$ on $\bar{C}^t$ is defined by

$$Z_C(d)(a_t) = \frac{t_s}{s_s} a_s \text{ for all } a \text{ in } \bar{C}, \quad Z_C(d)(\pm_t) = \pm_s,$$

$$Z_C(d)(\times_t) = \frac{s_s}{t_s} \times_s, \quad Z_C(d)(\div_t) = \frac{t_s}{s_s} \div_s, \quad Z_C(d)((a_t)^{*t}) = \frac{t_s}{s_s}(a_s)^{*s},$$

$$Z_C(d)(0_t) = 0_s, \quad Z_C(d)(1_t) = \frac{t_s}{s_s} 1_s. \tag{1.83}$$

The definition of the action of $Z_C(d)$ on complex conjugate numbers, shown in Eq. (1.83), can be obtained as follows. Let $a_t = (b + ic)_t$ with both b and c real. Then $(a_t)^{*t} = (b - ic)_t$. From this, one has

$$Z_C(d)((a_t)^{*t}) = Z_C(d)((b - ic)_t)$$
$$= \frac{t_s}{s_s}(b - ic)_s = \frac{t_s}{s_s}(a_s)^{*s}. \tag{1.84}$$

Note that

$$(a_t)^{*t} = (a^*)_t \text{ and } (a_s)^{*s} = (a^*)_s. \tag{1.85}$$

These equations show the distinction between $*$ and $*_t$. The complex conjugation $*$ applies to number values as in $a \to a^*$. $*_t$ applies to numbers as in $a_t \to (a_t)^{*t}$.

The demonstration that Z_C is a commutative group follows that for Z_R and will not be repeated here. One component of the proof not covered is that of transitivity for the complex conjugation operation. From Eq. (1.83), one has, with $s = dt$,

$$Z_C(e)Z_C(d)\left((a_t)^{*t}\right) = Z_C(e)\left(\frac{t_s}{s_s}(a_s)^{*s}\right) = Z_C(e)\left(\frac{t}{s}a^*\right)_s$$

$$= \frac{s_{es}}{(es)_{es}}\left(\frac{t}{s}a^*\right)_{es} = \frac{s_{es}}{(es)_{es}} \frac{t_{es}}{s_{es}}(a^*)_{es} \tag{1.86}$$

$$= Z_C(ed)(a^*)_t = Z_C(ed)\left((a_t)^{*t}\right).$$

It should be stressed that the maps $Z_C(d)$ are identity maps on the numbers in B_C. They change the representations of numbers in one structure to those in another structure.

The components of the structure $\bar{C}_s^t$ are obtained from Eq. (1.83). One has

$$\bar{C}_s^t = \left\{ B_C, \pm_s, \frac{s_s}{t_s} \times_s, \frac{t_s}{s_s} \div_s, \frac{t_s}{s_s}(a_s)^{*_s}, 0_s, \frac{t_s}{s_s} 1_s \right\}. \tag{1.87}$$

The representation of complex numbers in the form a_t, a_s is used here.

Note the order of scale change and complex conjugation in the description of $\bar{C}_s^t$. Scale change occurs after complex conjugation. Complex conjugation occurring after the scale change gives a different result in that $((t_s/s_s)a_s)^{*_s}$ is different from $(t_s/s_s)a_s^{*_s}$.

The difference between these two definitions can be seen by using the polar form for the complex scaling values. Let t and s be represented by $t = |t|e^{i\theta}$ and $s = |s|e^{i\phi}$. Then

$$\frac{t_s}{s_s}(a_s)^{*_s} = \left(\frac{t}{s}\right)_s a_s^{*_s} = \left(\frac{|t|}{|s|}e^{i(\theta-\phi)}a^*\right)_s \tag{1.88}$$

and

$$\left(\frac{t_s}{s_s}a_s\right)^{*_s} = \left(\frac{t}{s}a\right)_s^{*_s} = \left(\frac{|t|}{|s|}e^{-i(\theta-\phi)}(a)^*\right)_s. \tag{1.89}$$

The difference between the two is shown by the complex conjugation of the exponent in the scale factor in the second equation.

In the environment provided by the structure, $\bar{C}_s^t$, the number $(t_s/s_s)1_s$ is the identity. This follows from the fact that the number satisfies the axiom for a multiplicative identity. That is,

$$a_t \times_t 1_t = a_t \to Z_C(d)(a_t \times_t 1_t)$$

$$= Z_C(d)(a_t) \to Z_C(d)(a_t)Z_C(d)(\times_t)Z_C(d)1_t)$$

$$= Z_C(d)(a_t) \to \frac{t_s}{s_s}a_s\frac{s_s}{t_s} \times_s \frac{t_s}{s_s}1_s \tag{1.90}$$

$$= \frac{t_s}{s_s}a_s \to a_s \times_s 1_s = a_s.$$

This shows that if 1_t satisfies the multiplicative identity axiom in $\bar{C}^t$, so does $\frac{t_s}{s_s}1_s$ satisfy the axiom in $\bar{C}^t_s$ and 1_s satisfy the axiom in $\bar{C}^s$.

In addition, $\frac{t_s}{s_s}1_s$ is real in $\bar{C}^t_s$. This follows from

$$(1_t)^{*t} = 1_t \rightarrow Z_C(d)\big((1_t)^{*t}\big)$$
$$= Z_C(d)(1_t) \rightarrow \frac{t_s}{s_s}(1_s)^{*s}$$
$$= \frac{t_s}{s_s}1_s \rightarrow (1_s)^{*s} \tag{1.91}$$
$$= 1_s.$$

This result shows that if 1_t is real in $\bar{C}^t$, then $\frac{t_s}{s_s}1_s$ is real in $\bar{C}^t_s$ and 1_s is real in $\bar{C}^s$.

Most of the properties of real numbers under scaling apply to the complex numbers. For example, the action of $Z_C(d)$ on complex valued functions $f_t : B_{C^t} \rightarrow B_{C^t}$ is given by

$$Z_C(d)\big(f_t(a_t)\big) = \frac{t_s}{s_s}f_s(a_s). \tag{1.92}$$

The function $f_s : B_{C^s} \rightarrow B_{C^s}$ is a translation of f_t to the basis set for $\bar{C}^s$.

From Eq. (1.67), one has

$$Z_C(d)\int f_t(c)dc = \frac{t_s}{s_s}\int f_s(u)du. \tag{1.93}$$

Here $c = (t_s/s_s)u$. Equation (1.69) for derivatives is valid for complex numbers. It is repeated here as

$$Z_C(d)\left(\frac{df_t(c)}{dc}\right) = \frac{t_s}{s_s}\frac{df_s(u)}{du}. \tag{1.94}$$

This completes the details of the effect of the distinction between number and number value or meaning on the different types of numbers used in physics. Some additional comments are in order.

The scaling of structures for complex numbers provides a venue for the expansion of scaling for the real and rational numbers as well as the integers and natural numbers. It follows from the fact that these types of numbers can be described as substructures of the complex numbers. In particular, real and rational numbers are subfields of the complex numbers. The scaling of these other types of numbers can be extended to scaling with complex number values. For example, one can describe scaled natural number structures, $\bar{N}^t$, $\bar{N}^s$, and $\bar{N}^t_s$, where t and s are complex number values. Similar descriptions hold for $\overline{Ra}^t$, $\overline{Ra}^s$, and $\overline{Ra}^t_s$, and $\bar{R}^t_s$, $\bar{R}^s$, and $\bar{R}^t_s$.

A special case of this is based on the observation that natural numbers, integers, and rational numbers are substructures of the real numbers. Let r be a real scaling factor. The natural number structure is

$$\bar{N}^r = \{N_R, +_r, \times_r, <_r, 0_r, 1_r\}. \tag{1.95}$$

Here N_r is a subset of the real numbers where numbers in N_r have the form n_r. Also n denotes any natural number value.

Application of the number preserving value changing map to Z^r_1 gives

$$Z^r_1 \bar{N}^r = \bar{N}^r_1 = \left\{N_r, +, \frac{1}{r}\times, <, 0, r\right\}. \tag{1.96}$$

The representation n_r of the numbers in N_r becomes $n_r \to rn$. The subscript 1 has been suppressed on the components of $\bar{N}^r_1$ because of the equivalence between the numbers and the components $+, \times, <$ and their values or meaning.

An example of this has $r = \pi$. Then N_π is the set of all non-negative multiples of π. These numbers appear in phase factors, $e^{in\pi}$, used in physics.

Extension of the description for natural numbers to rational numbers is straightforward. Action of Z^r_1 on

$$\bar{Ra}^r = \{Ra_r, \pm_r, \times_r, \div_r, <_r, 0_r, 1_r\} \tag{1.97}$$

gives

$$Z^r_1 \bar{Ra}^r = \bar{Ra}^r_1 = \{Ra_r, \pm, \frac{1}{r}\times, r\div, <, 0, r\}. \tag{1.98}$$

The numbers s_r in Ra_r become $(rs)_1 = rs$ in $\bar{Ra}^1$. Here s is any rational number. Replacement of the subscript 1 by a rational number t gives

$$Z^r_t \bar{Ra}^r = \bar{Ra}^r_t = \left\{Ra_r, \pm_t, \frac{t}{r}\times_t, \frac{r}{t}\div_t, <_t, 0_t, \frac{r}{t}1_t\right\}. \tag{1.99}$$

The form of the numbers s_r in Ra_r changes to $[(r/t)s]_t$.

The type of number structure scaling described here is linear. More general nonlinear scalings and non-Diophantine arithmetic are described by Burgin and Czachor [56–61]. Nonlinear scalings are the basis of the formalism of non-Newtonian calculus introduced by Grossman and Katz [62–64], Maslov's idempotent analysis [65,66], and Pap's g-calculus [67–69].

Linear scaling or valuation has the advantage that it is easier to describe and use than are the more general scalings. Also it seems sufficient for the description of valuation as used here in physics and geometry. For this reason, the more general scalings are not used in this book.

It should be reemphasized that, with one exception, numbers, as elements of base sets B_N, B_{Ra}, B_R, and B_C, have no intrinsic value. Their value is determined by the structure containing them. Since they can belong to many different structures, each with different scale factors, they can have many different values.

The only exception is the number with value 0. To see this let $m = 0_r$. Then $m +_r b_r = 0_r +_r b_r = b_r$ shows that m is the additive identity in $\bar{S}^r$. The invariance of the value of m under change of scale factor from r to s follows from $0_r \to (r/s)_s 0_s = [(r/s)0]_s = 0_s$. The number has the usual name, 0. This is the only number whose value, 0, is independent of the structure containing it. The lack of factors of the number, 0, in the above is a consequence of this.

It is interesting to speculate on considering 0 to be the number vacuum. This notion is helped by the fact that its value is invariant under scaling. In analogy with the physical vacuum, one might consider positive numbers and negative numbers as numbers and "antinumbers" that annihilate each other. That is, for any number c, $c + -c = 0$.

1.8 Vector spaces

Number structure scaling affects vector spaces. This is a consequence of the fact that vector spaces are defined over a scalar field. The scalar fields are a number field of some type, such as the real or complex numbers. Because of their importance to physics, spaces based on complex number fields are discussed. Vector spaces based on real numbers are also important, especially in the application of scaling or number valuation to geometry. Consideration of these vector spaces will be deferred to later chapters where they will be much used.

The connection between vector spaces and complex number fields means that for each scaling number value, t, vector spaces of different dimensions have $\bar{C}^t$ as their scalar field. For each dimension, these spaces will be denoted by $\bar{V}^t$ with the dimension understood. Often the presence of $\bar{C}^t$ will be assumed without specific mention.

Here and from now on, the scaling factors will be restricted to be positive real numbers. This simplifies the discussion. Also the restriction is used for all applications of Part II.

Let s and t be two positive real number scaling values. The components of the normed vector spaces $\bar{V}^s$ and $\bar{V}^t$ are defined by

$$
\begin{aligned}
\bar{V}^t &= \{B_{V^t}, \pm_t, \odot_t, |\psi|_t, \psi_t\}, \\
\bar{V}^s &= \{B_{V^s}, \pm_s, \odot_s, |\phi|_s, \phi_s\}.
\end{aligned}
\tag{1.100}
$$

The same type of notation is used for the normed vector spaces as for the complex numbers. The components of the vector spaces are $\odot$ as scalar vector multiplication and $|-|_s$ and $|-|_t$ as norms. They are respective maps from vectors in $\bar{V}^s$ and $\bar{V}^t$ to non-negative real numbers in $\bar{C}^s$ and $\bar{C}^t$.

Note that $|\psi_s|$, as the norm of the vector, ψ, is the same number in $\bar{C}^s$ as is $|\psi|_s$. This is the number whose value is the norm $|\psi|$. Commutation between valuation and the norm operation is used to obtain this equality.

The base sets B_{V^t} and B_{V^s} contain all vectors in the form exemplified by ψ_t and ϕ_s. These vectors have vector values, ψ and ϕ, in their respective scaled vector spaces. Their numerical properties are based on numbers in the complex number spaces associated with the vector spaces. For example, the numerical properties of ϕ_s are based on numbers in $\bar{C}^s$. Norms of vectors and the coefficients of expansion of ϕ_s in basis vectors are numbers in $\bar{C}^s$. Numerical properties of ψ_t are based on numbers in $\bar{C}^t$.

Value maps act in a similar way to those for the complex number structures. The maps v_t and v_s are isomorphisms from $\bar{V}^t$ and $\bar{V}^s$

to $\bar{V}$. This is shown by

$$v_s(\gamma_s \pm_s \phi_s) = v_s(\gamma_s) \pm v_s(\phi_s) = \gamma \pm \phi,$$
$$v_s(a_s \odot_s \phi_s) = v_s(a_s) \odot v_s(\phi_s) = a \odot \phi, \qquad (1.101)$$
$$v_s(|\phi|_s) = |\phi|.$$

The value function can be extended to relate a vector α in the base set B_V to vectors in $\bar{V}^t$ and $\bar{V}^s$. One has

$$\alpha \to [v_t(\alpha)]_t = \psi_t \text{ and } \alpha \to [v_s(\alpha)]_s = \phi_s. \qquad (1.102)$$

These equations show that ψ is the vector value of α in $\bar{V}^t$ and ϕ is the vector value of α in $\bar{V}^s$.

The base sets B_{V^t} and B_{V^s} are defined using this extension. One has

$$B_{V^t} = \cup_{\alpha \epsilon B_V} [v_t(\alpha)]_t \qquad (1.103)$$

and

$$B_{V^s} = \cup_{\alpha \epsilon B_V} [v_s(\alpha)]_s. \qquad (1.104)$$

The definition for v_t acting on $\bar{V}^t$ is obtained from Eq. (1.101) by replacing s with t. As was the case for numbers, there is a vector, ψ_s, in $\bar{V}^s$ that has the same vector value as does ψ_t in $\bar{V}^t$. The two vectors, ψ_s and ψ_t, are different from one another. It follows that there is another vector, β in the base set, different from α such that

$$\beta \to [v_s(\beta)]_s = \psi_s.$$

The relation between ψ_s and ψ_t is assumed to be similar to that for numbers in that

$$sv_s(\psi_t) = tv_t(\psi_t) = tv_s(\psi_s) = t\psi. \qquad (1.105)$$

From Eq. (1.105), one obtains

$$sv_s(\psi_t) = v_s(s_s\psi_t) = tv_s(\psi_s) = v_s(t_s\psi_s). \qquad (1.106)$$

This equation yields

$$\psi_t \equiv \frac{t_s}{s_s}\psi_s = \left[\frac{t}{s}\psi\right]_s. \qquad (1.107)$$

Support for this relation, up to a phase factor, can be obtained from the relation between the norms of the vectors. From Eq. (1.76), one has

$$|\psi_t|^t = |\psi|_t = \frac{t_s}{s_s}|\psi|_s = \frac{t_s}{s_s}|\psi_s|^s. \tag{1.108}$$

This equation gives

$$\psi_t = e^{i\phi}\frac{t_s}{s_s}\psi_s. \tag{1.109}$$

Here $e^{i\phi}$ is an arbitrary phase factor.

Equation (1.107) shows that ψ_t is the same vector in $\bar{V}^t$ as $[\frac{t}{s}\psi]_s$ is in $\bar{V}^s$. Their vector values differ by t/s. The equation can also be interpreted as showing that the effect of a vector preserving value changing map of ψ_t to a vector in $\bar{V}^s$ is to multiply ψ_s by a scaling factor.

Extension of the restriction of the value function for numbers to scalar properties of vectors for a unity scaling factor means that the number representing the property is identified with the value of the number. The norm is an example. If $v_1(\beta) = \psi$, then $|v_1(\beta)|_1 \equiv |v_1(\beta)|$ or $|\psi|_1 \equiv |\psi|$.

The same restriction, applied to vectors in $\bar{V}^1$, gives $v_1(\beta) \equiv \beta$ for any vector, β. If $v_1(\beta) \equiv \psi$, then the restriction gives $\psi_1 \equiv \psi$.

As was the case for numbers, one can define two groups of maps: One group, W_V, consists of vector changing and value preserving maps, such as $W_V(d)$. The other group, Z_V, contains vector preserving value changing maps, $Z_V(d)$. The parameter, d, is a real or complex number value. The type depends on the scalars associated with the vector space. The map, $W_V(d)$, extends the definition for complex numbers to include vector spaces. It is defined by

$$W_V(d)\bar{V}^t = \bar{V}^s, \tag{1.110}$$

where

$$W_V(d)(\psi_t) = \psi_s, \quad W_V(d)(\pm_t) = \pm_s, \quad W_V(d)(\odot_t) = \odot_s,$$
$$W_v(d)(c_t \odot_t \psi_t) = c_s \odot_s \psi_s, \quad W_V(d)(|\psi|_t) = |\psi|_s. \tag{1.111}$$

As before, $s = dt$ and c_t is any complex number in $\bar{C}^t$.

The other map type differs from $W_V(d)$ in that it preserves vectors but not vector values. The definition is similar to that for complex numbers in Eq. (1.83). One has with $s = dt$

$$Z_V(d)(\bar{V}^t \times \bar{C}^t) = \bar{V}_s^t \times \bar{C}_s^t. \tag{1.112}$$

Complex number structures are included here because of the dependence of vector spaces on the underlying scalar field. The map shows explicitly that just as $\bar{C}^t$ is the scalar structure for $\bar{V}^t$, so is $\bar{C}_s^t$ the scalar structure for $\bar{V}_s^t$.

The actions of $Z_V(d)$ on the components of $\bar{C}^t$ and the components of $\bar{C}_s^t$ are given by Eqs. (1.83) and (1.87), so they will not be repeated here. The actions of $Z_V(d)$ on the components of $\bar{V}^t$ are given by

$$Z_V(d)(\psi_t) = \frac{t_s}{s_s}\psi_s, \;\; Z_V(d)(\pm_t) = \pm_s, \;\; Z_V(d)(\odot_t) = \frac{s_s}{t_s}\odot_s,$$

$$Z_V(d)(|\psi|_t) = \frac{t_s}{s_s}|\psi|_s. \tag{1.113}$$

As was the case for complex numbers, $Z_v(d)$ implements a vector structure environment change in that vectors and operations in $\bar{V}^t$ are given their expressions in terms of vectors and operations in $\bar{V}^s$.

The components of $\bar{V}_s^t$ are those in Eq. (1.113). They are shown in

$$\bar{V}_s^t = \left\{ B_V, \pm_s, \frac{s_s}{t_s}\odot_s, \frac{t_s}{s_s}|\psi|_s, \frac{t_s}{s_s}\psi_s \right\}. \tag{1.114}$$

These definitions of $\bar{V}_s^t$ express the components of $\bar{V}^t$ in terms of those of $\bar{V}^s$. In this equation, $\frac{t_s}{s_s}|\psi|_s = (\frac{t}{s}|\psi|)_s$.

Scaling factors are present for scalar vector multiplication. If $c_t = (t_s/s_s)c_s$ is a complex number in $\bar{C}_s^t$ (Eq. (1.87)), then Eq. (1.113) gives

$$Z_V(d)(c_t \odot_t \psi_t) = Z_V(d)(c_t)Z_V(d)(\odot_t)Z_V(d)(\psi_t)$$

$$= \frac{t_s}{s_s}c_s \frac{s_s}{t_s}\odot_s \frac{t_s}{s_s}\psi_s = \frac{t_s}{s_s}c_s \odot_s \psi_s. \tag{1.115}$$

This equation shows that scaling of the $\odot$ operation is the same as that for multiplication in scalars. It is needed so that the representation of a vector as the product of a scalar and another vector scales by the same factor as does either the scalar or the vector in the scalar vector product.

1.8.1 *Hilbert spaces*

Hilbert spaces are vector spaces in which the norm is replaced by an inner product. The value structure $\bar{H}$ and scaled structures, $\bar{H}^t$ and $\bar{H}^s$, are given by

$$\bar{H} = \{H, \pm, \odot, \langle \phi, \psi \rangle, \psi\},$$

$$\bar{H}^t = \{B_H, \pm_t, \odot_t, \langle \phi_t, \psi_t \rangle, \psi_t\}, \tag{1.116}$$

$$\bar{H}^s = \{B_H, \pm_s, \odot_s, \langle \phi_s, \psi_s \rangle, \psi_s\}.$$

The set B_H is the common set of Hilbert space vectors, and $\bar{H}$ is the set of vector values. The vectors in $\bar{H}^t$ and $\bar{H}^s$ with values ψ and ϕ in $\bar{H}$ are denoted by ψ_t, ϕ_t and ψ_s, ϕ_s. The scalar products $\langle \phi_s, \psi_s \rangle$ and $\langle \phi_t, \psi_t \rangle$ map pairs of vectors to numbers in $\bar{C}^s$ and $\bar{C}^t$. Note that $\langle \phi_s, \psi_s \rangle = \langle \phi, \psi \rangle_s$ and $\langle \phi_t, \psi_t \rangle = \langle \phi, \psi \rangle_t$.

The action of the vector changing value preserving map, $W_V(d)$ on $\bar{H}^t$, satisfies

$$W_V(d)\bar{H}^t = \bar{H}^s, \tag{1.117}$$

where $s = dt$. The map components are given by Eq. (1.111) except that $W_V(d)|\psi_t| = |\psi_s|$ is replaced by $W_V(d)\langle \phi_t, \psi_t \rangle = \langle \phi_s, \psi_s \rangle$.

The action of the vector preserving value changing map, $Z_V(d)$ on $\bar{H}^t$, is given by

$$Z_V(d)\bar{H}^t = \bar{H}^t_s, \tag{1.118}$$

where

$$\bar{H}^t_s = \left\{ B_H, \pm_s, \frac{s_s}{t_s}\odot_s, \frac{t_s}{s_s}\langle \phi_s, \psi_s \rangle, (t_s/s_s)\psi_s \right\}. \tag{1.119}$$

The relation, $Z_V(d)\langle \phi_t, \psi_t \rangle = (t_s/s_s)\langle \phi_s, \psi_s \rangle$, describes the scaling resulting from the change in environment on the scalar product. It is essentially the same change as that for the norm in vector spaces in general. It is quite different than the term $\langle Z_v(d)\phi_t, Z_V(d)\psi_t \rangle$. One has

$$\langle Z_v(d)\phi_t, Z_V(d)\psi_t \rangle = \left\langle \frac{t_s}{s_s}\phi_s, \frac{t_s}{s_s}\psi_s \right\rangle = \left(\frac{t_s}{s_s}\right)^{*_s} \frac{t_s}{s_s}\langle \phi_s, \psi_s \rangle. \quad (1.120)$$

This is another example of the fact that the order of carrying out operations matters in the presence of scaling. One sees here that

$$Z_V(d)\langle \phi_t, \psi_t \rangle) \neq \langle Z_v(d)\phi_t, Z_V(d)\psi_t \rangle. \quad (1.121)$$

This inequality becomes an equality in the absence of scaling or in the special case where $t_s/s_s = 1_s$.

Vector spaces are examples of types of mathematical systems different from numbers that are affected by number structure scaling. Many other types of systems are also affected by scaling. Any type of system that includes numbers as part of their axiomatic description is included. Algebraic systems that are closed under multiplication by scalars are included, so are group representations as matrices of numbers.

So far there has been no discussion on how these scaled mathematical structures are to be used in physics and geometry. Their use is based on one of the basic assumptions behind the material in this book, that mathematical structures are local instead of being global (outside of space and time) as is the usual assumption. Local mathematics means that separate mathematical structures are associated with each point of a manifold, M, taken here to be space, time, or space–time.

Structures localized at a point of M are denoted with a location subscript. For example, a real number structure scaled by the number t and localized at point x is denoted by $\bar{R}_x^t$. A vector space structure scaled by the number s and localized at point y is denoted by $\bar{V}_y^s$. How these structures are to be used in physics and geometry is described in the following chapter.

Appendix

The proof that W_R is a commutative group follows from

$$W_R(e)W_R(d)\bar{R}^t = W_R(e)\bar{R}^s = \bar{R}^{edt} = W_R(ed)\bar{R}^t. \quad (1.122)$$

From this, one has

$$W_R(d)W_R(e) = W_R(e)W_R(d) \text{ and } W_R(d)^{-1} = W_R(d^{-1}). \quad (1.123)$$

$W_R(1)$ is the identity.

The proof for Z_R is less straightforward. One has, with no substitution $s = dt$,

$$Z_R(e)Z_R(d)\bar{R}^t = Z_R(e)\bar{R}^t_{dt}. \quad (1.124)$$

The actions of $Z_R(e)$ on the numbers and basic operations of $\bar{R}^t_{dt}$ are given by

$$Z_R(e)Z_R(d)(a_t) = Z_R(e)\left(\left(\frac{t}{dt}a\right)_{dt}\right) = \frac{dt_{edt}}{edt_{edt}}\left(\frac{t}{dt}a\right)_{edt}$$

$$= \frac{t_{edt}}{edt_{edt}}a_{edt} = Z_R(ed)(a_t),$$

$$Z_R(e)(\pm_{dt}) = \pm_{edt},$$

$$Z_R(e)\left(\frac{t_{dt}}{dt_{dt}} \div_{dt}\right) = Z_R(e)\left(\left(\frac{t}{dt}\div\right)_{dt}\right) = \frac{dt_{edt}}{edt_{edt}}\left(\frac{t}{dt}\div\right)_{edt}$$

$$= \frac{t_{edt}}{edt_{edt}}\div_{edt} = Z_R(ed)(\div_t),$$

$$Z_R(e)\left(\frac{dt_{dt}}{t_{dt}} \times_{dt}\right) = Z_R(e)\left(\left(\frac{dt}{t}\times\right)_{dt}\right) = \frac{edt_{edt}}{dt_{edt}}\left(\frac{dt}{t}\times_{\bullet}\right)_{edt}$$

$$= \frac{edt_{edt}}{t_{edt}}\times_{edt} = Z_R(ed)(\times_t). \quad (1.125)$$

These equations make use of the fact that $((t/dt)a)_{dt}$ in $\bar{R}^t_{dt}$ and $((t/dt)a)_{edt}$ in $\bar{R}^t_{edt}$ have the same $\bar{R}$ value. That is, $v_{dt}(((t/dt)a)_{dt}) = v_{edt}(((t/dt)a)_{edt})$. However, they are different numbers. The relation between them is given by Eq. (1.47). The equations use the fact that the same number has different expressions in different real number structures. The same arguments are used for the action of Z_R on multiplication and division. $\div_{dt}$ and $\times_{dt}$ are the same operations in $\bar{R}^{dt}$ as $\div_{edt}$ and $\times_{edt}$ are in $\bar{R}^{edt}$. However they are different operations in that they are related by scaling factors. Since Z_R must also preserve operations, the effect of Z_R is to express $\div_{dt}$ and $\times_{dt}$ in terms of $\div_{edt}$ and $\times_{edt}$. This is done by the relations $\div_{dt} = (dt_{edt}/edt_{edt}) \div_{edt}$ and $\times_{dt} = (edt_{edt}/dt_{edt}) \times_{edt}$.

Commutativity and the existence of an inverse in the group Z_R follow from noting that $Z_R(ed) = Z_R(de)$ and from setting $e = d^{-1}$. A simple expression of the action of the Z_R comes from noting that $\bar{R}^t = \bar{R}^t_t$. Since $Z_R(d)\bar{R}^t_t = \bar{R}^t_{dt}$, one can use Eq. (1.125) to extend this to

$$Z_R(d)\bar{R}^t_s = \bar{R}^t_{ds}. \qquad (1.126)$$

This holds for any number values, t, d, s.

Chapter 2

Fiber Bundles and Connections

2.1 Fiber bundles

Fiber bundles provide a very good framework for the use of local mathematical structures in physics and geometry [26]. In essence, they provide a suitable mathematical arena for the discussion. This use is in addition to their use in many areas of physics and geometry. These include gauge theories [27,28], quantum mechanics [36–40], and geometry [35]. Here they will be used to apply number structure scaling or valuation to gauge theories, quantum mechanics, and geometry.

A fiber bundle [41,42] can be described as a triple, $\mathfrak{B} = \{E, \pi, M\}$. Here E and M are the total and base spaces and π is a projection of E onto M. For each point, x, in M, $\pi^{-1}(x)$ is a fiber, at x, in E.

A fiber bundle is a product or trivial bundle if the total space is a product as in $E = M \times F$. Here F is a fiber and $\pi^{-1}(x) = (x, F) = F_x$ is the fiber at x in M. In this work, the base space, M, is flat as it is limited to Euclidean or Minkowski spaces.

A hair brush is a simple home example of a fiber bundle [43]. The bristles are the fibers at different locations on the base. A simple standard example of a nontrivial fiber bundle is the set of line segments glued to a Mobius strip [43]. Locally, the line segments form part of a cylinder, but globally, there is a twist present.

Fibers can contain many different types of mathematical systems. A simple example is the fiber product [41] of the trivial bundles

$M \times \bar{C}^r, \pi, M$ and $M \times \bar{V}^r, \pi, M$. This is given by

$$\mathfrak{CV} = M \times F, \pi, M. \tag{2.1}$$

Here the fiber $F = \bar{C}^r \times \bar{V}^r$ contains the complex number structure, $\bar{C}^r$, and the vector space structure, $\bar{V}^r$. The scaling factor is r. The projection operator π is defined by $\pi(x \times F) = x$. The inverse of the projection operator defines the fiber at x as in

$$\pi^{-1}(x) = x \times F = F_x = \bar{C}_x^r \times \bar{V}_x^r. \tag{2.2}$$

Here $\bar{C}_x^r$ and $\bar{V}_x^r$ are a complex number structure and a vector space structure at point x of M. They are local mathematical structures in that they are localized to a point of M. For vector spaces over real numbers, $\bar{C}_x^r$ is replaced by $\bar{R}_x^r$. The superscript, r, is restricted to be a positive real number valued scaling or value factor.

The contents of the fiber shown above are an example of the possible contents of a fiber. In general, the contents of a fiber depend on the problem being considered. For gauge theories, the vector spaces are low-dimensional spaces. This is the case for the standard model where the spaces account for the internal degrees of freedom for the elementary particles. For quantum mechanical properties, such as those described by wave functions, the vector space is a Hilbert space of infinite dimensionality. For geometries, the space depends on the problem under consideration. The associated number structure is that for real numbers.

The bundle fiber can be expanded to include structures with more than one scaling factor. A simple case is a fiber that contains real number structures for two positive, real scaling or value factors, s and t. For each point, x in M, the fiber F_x is given by[1]

$$F_x = \bar{R}_x^s + \bar{R}_x^t. \tag{2.3}$$

The base sets of elements, $B_{R_x^s}$ and $B_{R_x^t}$, are defined in Eq. (1.43). The fiber bundle is a sum of the two bundles, $M \times \bar{R}^s, \pi, M$ and $M \times \bar{R}^t, \pi, M$.

---

[1]The definition can be expanded to include negative values of s or t. In this case, the direction of the order relation may have to be changed.

The fibers can also contain products of structures. An example is the inclusion of vector spaces with the associated scalars in the fiber F_x where

$$F_x = \bar{R}_x^s \times \bar{V}_x^s + \bar{R}_x^t \times \bar{V}_x^t. \tag{2.4}$$

The description of fibers in Eq. (2.4) has been described for a very simple fiber structure. The fibers and fiber bundle extension of these maps to more complex fibers and fiber bundles are based on taking into account the freedom of choice of values for numbers and vector spaces at different locations of M.

The freedom to choose values for numbers and vectors at different locations in M means that the values of s and t can depend on locations in M. This is accounted for here by the introduction of a smooth real valued field, g. The number value, $g(x)$, is the scaling or value factor for the complex number and vector space structures associated with point x of M. The value field g will also be used to give the scaling or value factors for the fiber contents as real or complex numbers and vector spaces with different properties. If needed, this can be extended to other types of mathematical systems whose description includes numbers.

There are an infinite number of possibilities for the field g. It can be a constant field. Or it can vary rapidly or slowly over M. An argument for allowing g to vary with location is the extension of the argument of Yang and Mills [25], for vectors in isospin space to numbers. The extension, stated earlier in Chapter 1, is that the value of a number at one location does not determine the value of the same number at another location.

The fiber bundle for complex numbers and vector spaces, with the g field taken into account, is given by

$$\mathfrak{CV}^g = E, \pi, M. \tag{2.5}$$

This is a nontrivial fiber bundle. The total space, E, cannot be separated into the product, $M \times F$, because there is no global representation of F.

The usual method of describing this bundle [44] is to decompose it into a collection of product fiber bundles, each defined on an open subset or region of M. The total space for each of these bundles is expressed by

$$E = U \times F^{g(U)}. \tag{2.6}$$

The size of the regions depends on the variation of g. Each region, U, should be small enough so that the variation of g over the points of U is quite small. In this case, the fiber $F^{g(U)}$ for complex numbers and vector spaces becomes

$$F^{g(U)} = \bar{C}^{g(U)} \times \bar{V}^{g(U)}. \tag{2.7}$$

The fibers in $U \times F^{g(U)}$ are defined by

$$\pi_U^{-1}(x) = F_x^{g(U)} = \bar{C}_x^{g(U)} \times \bar{V}_x^{g(U)}. \tag{2.8}$$

Here x is any point in U.

The number value $g(U)$ can be defined in different ways. It can be a representative value of $g(x)$ for all x in U or it can be an average of the values of $g(x)$ over U. At this point, the choice is arbitrary.

The bundles in the collection must satisfy a consistency condition [44]. This is the requirement that if U and U' overlap in M, then the transition from $F^{g(U)}$ to $F^{g(U')}$ in the overlap region should be smooth. This can be expressed by the requirement that if $\pi_U^{-1}(U) = F^{g(U)} \times U$ and $\pi_{U'}^{-1}(U') = F^{g(U')} \times U'$, there should be a smooth transition function ϕ such that

$$\phi(\pi_U^{-1}(U \cap U')) = \pi_{U'}^{-1}(U \cap U'). \tag{2.9}$$

There is another way to describe a fiber bundle that is better suited to this work. In this description, the total space is represented as a product of M and a union of fibers for all real scaling factors. One has

$$E = M \times \cup_t F^t. \tag{2.10}$$

Here F^t denotes a generic fiber in which the mathematical structures in F^t are all scaled by t.

The contents of F^t can be adjusted to suit the problem under consideration. They can be very extensive consisting of many different mathematical structures or they can be limited to a very few structures. An example already considered is the case where

$$F^t = \bar{C}^t \times \bar{V}^t. \tag{2.11}$$

The effect of the g field is accounted for by changing the definition of the projection operator to be a g-dependent projection operator, π_g. The domain of π_g on E is all of M and the subset of fibers whose scaling factors, t, are in the range of values taken by g. If the range set of g is all possible positive real numbers, then the domain of π_g is all of E. If the values of g are constrained to lie in an interval, I, of real numbers, then the domain of π_g is restricted to the set of all $M \times F^t$, where t is in I.

The fiber bundle with the g field included and π_g as the projection operator is given by

$$\mathfrak{M}\mathfrak{F}^g = M \times F, \pi_g, M, \tag{2.12}$$

where

$$F = \cup_t F^t. \tag{2.13}$$

The projection operator π_g is defined so that the inverse selects the fiber with the scaling factor t given by the value of g at the location under consideration. That is,

$$\pi_g^{-1}(x) = F_x^{g(x)}. \tag{2.14}$$

This method of defining F as the union of fibers for all t values and π_g as a specific fiber selection operator allows for flexibility in setting the contents of all the F^t. In this way, the contents of all the F^t can be altered to suit the problem at hand without changing the structure of the fiber bundle.

The fiber bundle, $\mathfrak{M}\mathfrak{F}^g$, provides a good mathematical arena for discussing the effects of the value field and localization of mathematical structures on physical and geometric quantities. The fiber contents depend on the type of problem under consideration. For example, the vector space can be finite or infinite dimensional. It can be a Hilbert space, or it can be the direct or tensor product of different vector spaces. For geometric properties,

$$F_x^{g(x)} = \bar{R}_x^{g(x)} \times \mathbb{S}_x^{g(x)}. \tag{2.15}$$

Here $\mathbb{S}_x^{g(x)}$ is a local image space of M scaled by $g(x)$.

Both physical and geometric properties can be described with a single fiber bundle if F^t is expanded to contain, $\bar{C}^t, \bar{V}^t, \bar{R}^t$, and $\mathbb{S}^t$. In this case,

$$F_x^{g(x)} = \bar{C}_x^{g(x)} \times \bar{V}_x^{g(x)} \times \bar{R}_x^{g(x)} \times \mathbb{S}_x^{g(x)}. \tag{2.16}$$

Structures for other mathematical systems can be added to each F^t if desired.

2.2   Values of numbers at different locations

The application of the foregoing description of fiber bundles to problems in physics and geometry requires a method to assign values to numbers at different locations. As an example let h be a number at location y in M. By itself the number h has no intrinsic value. The number 2.667 is a string of five symbols. The arrangement of the symbols shows that it is a rational number. By itself it has no specific value. Any value is possible.

The value of this number is determined by the location of the number structure associated with h. If $\bar{S}_x^{g(x)}$ is a number structure at x, then the value of h in $\bar{S}_x^{g(x)}$ depends in general on the locations, y and x, and the values, $g(y)$ and $g(x)$, of the value function g.

The method used here to determine the value depends on the value function, v, described earlier. A useful way to proceed consists of first associating a number at a location with the number structure at the same location. The number h at y is associated with the number structure $\bar{S}_y^{g(y)}$. The value of this number is $v_{g(y)}(h)$. The associated number is $[v_{g(y)}(h)]_{g(y)}$. Here S denotes a number type, rational, real, or complex that is suitable for the problem at hand.

Determination of the value of h at another location x is implemented by a parallel transport operation. This operation determines the value of h at y in the number structure at x.[2] Implementation of parallel transport operations, as connections or number preserving value changing operations, is described in Section 2.3.

2.2.1 *Fields on* $\mathfrak{M}\mathfrak{F}^g$

Many physical and geometric properties are described by means of fields whose domain is all or part of M as space or space–time. Each field assigns a number or a vector to each point of M. The usual description makes no distinction between number and number value and vector and vector value at each location in M. As a result a field of numbers is identified with a field of number values. A field of vectors is identified with a field of vector values.

In the context of local mathematical structures and the presence of the value field, g, the concepts of number and vector are distinct from number value and vector value. It follows that number fields are different from number value fields. Vector fields are different from vector value fields.

To see the distinction, let α be a number field. For each x in M, $\alpha(x)$ is a number in the base set, B_S. Here S stands for rational, real, or complex numbers, whichever is appropriate for the problem being considered.[3]

A priori the field numbers $\alpha(x)$ are meaningless. They have no intrinsic value. A natural association between number field elements and number structures fixes this problem. The fix consists of an association of the number $\alpha(x)$ with the colocated structure, $\bar{S}_x^{g(x)}$. The corresponding number value at x of the number value field, ψ, is

---

[2]This transport is not a physical motion of the number. It is just a change of the location of the mathematical number structure to which the number is associated.

[3]Recall that B_S is the set of all numbers of type S. The numbers have no specific values or location. Any value of a number or location of the set is possible. Association of B_S to a location x in M results in assigning values to the set numbers. The set B_S becomes the base set, $B_{Sg(x)}$, of numbers in $\bar{S}_x^{g(x)}$, where

$$B_{Sg(x)} = \cup_{h \epsilon B_S} [v_{g(x)}(h)]_{g(x)}.$$

defined by

$$\psi(x) = v_{g(x)}(\alpha(x)). \tag{2.17}$$

The corresponding number field is defined by

$$[\psi(x)]_{g(x)} = [v_{g(x)}(\alpha(x))]_{g(x)}. \tag{2.18}$$

From this one sees that the meaningless number $\alpha(x)$ becomes the meaningful number, $[\psi(x)]_{g(x)}$. This is a number in the base set $B_{Sg(x)}$ for the number structure at x. For fiber bundles, this association for all x is described as lifting the field, α, to be a section [41] on the fiber bundle. This lift can be expressed by

$$\alpha(-) \to [\psi(-)]_{g(-)} = [v_{g(-)}(\alpha(-))]_{g(-)}. \tag{2.19}$$

The description for vector fields is similar to that for number fields. For each x, $\vec{\beta}(x)$ is a vector in $\bar{V}_x^{g(x)}$. The corresponding vector value field is defined by $\vec{\phi}(x) = v_{g(x)}(\vec{\beta}(x))$.

Figure 2.1 shows the relations between α and $\vec{\beta}$ as number and vector fields in M and their collocated lifted representations as number value and vector value fields in $\bar{C}_x^{g(x)}$ and $\bar{V}_x^{g(x)}$ with specified number values. The relations are shown for two different locations, x and y, in M.

Throughout this book, subscripts for numbers and vectors at different locations in M will be value factors rather than locations in M. For example, the number with value $\psi(x)$ at x is denoted by $\psi(x)_{g(x)}$ instead of $\psi(x)_x$. This is done to be consistent with the labeling of the components of different types of number and vector space structures described in the previous chapter.

2.2.2   *Value function choices*

A priori there are many possibilities for the functional dependence of $v_{g(x)}$ on $g(x)$. Part III will give a general description of the effect that freedom of choice of number values has on physics and geometry. This will be done without specifying the dependence of the value function on $g(x)$.

Part II shows the effect on physics and geometry of a specific choice of the value function. The description in the previous chapter

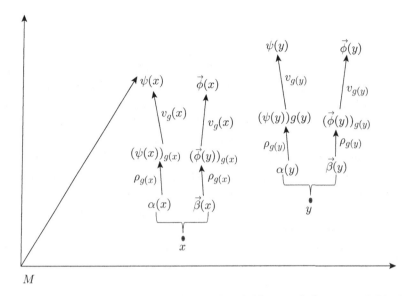

Figure 2.1. Representations of the number field, α, and the vector field, β, at two locations, x and y, and their lifts to elements of number and vector structures at x and y are shown. The lifts of these fields by chart maps, $\rho_{g(x)}$ and $\rho_{g(y)}$, to a number in $\bar{S}_x^{g(x)}$ and a vector in $\bar{V}_x^{g(x)}$ are included. Use of the valuation functions $v_{g(x)}$ and $v_{g(y)}$ gives an alternate form for these numbers and vectors as $[v_{g(x)}(\alpha(x))]_{g(x)} = \psi(x)_{g(x)}$ and $[v_{g(y)}(\alpha(y))]_{g(y)} = \psi(y)_{g(y)}$. Similar expressions hold for the vector field, β, at x and y.

of scaling and the relations between numbers and their values is used to determine the functional dependence of $v_{g(x)}$ on $g(x)$. The value function is restricted to satisfy the condition where for each location x, if $g(x) = 1$, then $v_1(h) = h$. The number, h, and its value are conflated. This was seen explicitly for rational and natural numbers in the previous chapter, Section 1.5 and Eq. (1.4).

This choice determines the functional dependence of $v_{g(x)}$ on $g(x)$ for all locations. One has

$$v_{g(x)}(h) = \frac{h}{g(x)}. \qquad (2.20)$$

This equation has the required property that $v_1(h) = h/1 = h$. It also shows that, contrary to what is usually assumed, *the value of a number depends on its location*. The dependence is given by the dependence of $g(x)$ on x. This location dependence is basic for all that follows in this work.

2.3 Connections

There are many situations in which one wants to compare or combine numbers or vectors in fibers at different locations. Examples include fields on M that are lifted to become sections over $\mathfrak{MF}^g$. The values of the fields at different locations depend on the value function that is different for different locations in M. Most of the discussion of connections will apply to the general function in which $\psi(y)$ is the value of the number $\alpha(y)$ at y and $\vec{\phi}(y)$ is the value of the vector $\vec{\beta}(y)$ at y. The special case where the value function is defined so that number and number value, and vector and vector value are conflated for the identity scaling factor is discussed at the ends of the descriptions for numbers and vectors.

Let α be a number field on M. The lift of the number field to a section on the fiber bundle gives

$$[v_{g(y)}(\alpha(y))]_{g(y)} = \psi(y)_{g(y)} \quad \text{and} \quad [v_{g(x)}(\alpha(x))]_{g(x)} = \psi(x)_{g(x)} \quad (2.21)$$

as numbers in $\bar{S}_y^{g(y)}$ and $\bar{S}_x^{g(x)}$ with number values, $\psi(y)$ and $\psi(x)$. As before S denotes rational, real, or complex numbers.

Theoretical descriptions of many physical and geometric properties include arithmetic combinations of field components at different locations. These combinations are not defined because the components are in number structures at different locations. Examples include addition or subtractions of field components at different locations.

This problem is solved by the use of connections to parallel transport numbers or vectors in structures at different locations to number or vector structures at a common location. These transports represent or define the notion of "sameness". A number or vector in a number or vector structure at one location, parallel transported to a number or vector in a number or vector structure at another location, is the same number or vector in the collocated structure as is the original number or vector in the collocated structure. The values of the number or vector can change under parallel transport.

Connections are extensions of the number preserving value changing and number changing value preserving maps between structure with different scaling factors as defined in Section 1.3. Applied to local number structures and vector spaces at y the maps give

$$Z_{g(x)}^{g(y)}(\bar{S}_y^{g(y)} \times \bar{V}_y^{g(y)}) = \bar{S}_{x,g(x)}^{g(y)} \times \bar{V}_{x,g(x)}^{g(y)} \qquad (2.22)$$

and

$$W_t^s(\bar{S}_x^s \times \bar{V}_x^s) = \bar{S}_x^t \times \bar{V}_x^t. \qquad (2.23)$$

Here $\bar{S}_{x,g(x)}^{g(y)}$ and $\bar{V}_{x,g(x)}^{g(y)}$ are given by Eqs. (1.87) and (1.114).

These extensions expand number preserving value changing maps between structures of the same mathematical type at the same location to structures that are at different locations in M. They are used here to map number or vector field values in different structures at different locations to number or vector structures at a single reference location. The values can then be combined as the combining operations such as addition or subtraction are defined within the structure.

The lifts of the number field α and vector field, $\vec{\beta}$, to number fields, $\psi(-)_{g(-)}$ and $\vec{\phi}(-)_{g(-)}$, on the number and vector spaces at each location on M provide good examples to illustrate and define the action of the connection operation. As noted these are number preserving value changing maps.

Let y and x be two locations on M. Define $C_g(x, y)$ to be the connection that maps $\psi(y)_{g(y)}$ in $\bar{S}_y^{g(y)}$ to the same number in $\bar{S}_x^{g(x)}$. The action of the connection is given by

$$\psi(y)_{g(y)} \to C_g(x, y)(\psi(y)_{g(y)}) = \frac{g(y)_{g(x)}}{g(x)_{g(x)}}(\psi(y)_{g(x)})$$

$$= \left(\frac{g(y)}{g(x)} \psi(y) \right)_{g(x)}. \qquad (2.24)$$

Here $\psi(y)_{g(x)}$ is the number in $\bar{S}_x^{g(x)}$ that has the same value, $\psi(y)$, as the number $\psi(y)_{g(y)}$ has in $\bar{S}_y^{g(y)}$. They are different numbers.

The difference between the two numbers is accounted for by the ratio

$$g(y)_{g(x)}/g(x)_{g(x)} = (g(y)/g(x))_{g(x)}.$$

Replacement of $\psi(y)$ by its equivalent, $v_{g(y)}(\alpha(y))$, in the definition of the connection gives

$$C_g(x,y)\left([v_{g(y)}(\alpha(y))]_{g(y)}\right) = \frac{g(y)_{g(x)}}{g(x)_{g(x)}}\left([v_{g(y)}(\alpha(y))]_{g(x)}\right)$$

$$= \left[\frac{g(y)}{g(x)}v_{g(y)}(\alpha(y))\right]_{g(x)}. \qquad (2.25)$$

In this equation, $[v_{g(y)}(\alpha(y))]_{g(x)}$ is the number in the number structure at x that has the same value as does the number $\alpha(y)$ in the number structure at y. As expected $[v_{g(y)}(\alpha(y))]_{g(y)}$ and $[v_{g(y)}(\alpha(y))]_{g(x)}$ are different numbers.

The last equalities in Eqs. (2.24) and (2.25) are based on the fact that the value of the product of two numbers in the same structure is equal to the product of the values of the two numbers. That is,

$$v_{g(x)}\left(\frac{g(y)_{g(x)}}{g(x)_{g(x)}}\psi(y)_{g(x)}\right) = v_{g(x)}\left(\left[\frac{g(y)}{g(x)}\psi(y)\right]_{g(x)}\right) = \frac{g(y)}{g(x)}\psi(y).$$

$$(2.26)$$

Equation (2.24) shows that $\left(\frac{g(y)}{g(x)}\psi(y)\right)_{g(x)}$ is the same number in $\bar{S}_x^{g(x)}$ as $\psi(y)_{g(y)}$ is in $\bar{S}_y^{g(y)}$. Their values differ by the factor $g(y)/g(x)$. The number $\psi(y)_{g(x)}$ has the same value, $\psi(y)$, as does $\psi(y)_{g(y)}$. But $\psi(y)_{g(x)}$ is not the same number as is $\psi(y)_{g(y)}$.

Figure 2.2 illustrates the effect of the connection as a number preserving value changing map for two locations, y and x in M. The figure shows that $\psi(y)_{g(y)}$ is the same number in $\bar{S}_y^{g(y)}$ as

$$(g(y)_{g(x)}/g(x)_{g(x)})\psi(y)_{g(x)}$$

is in $\bar{S}_x^{g(x)}$.

The connection can be defined from the number preserving value changing and number changing value preserving maps defined in

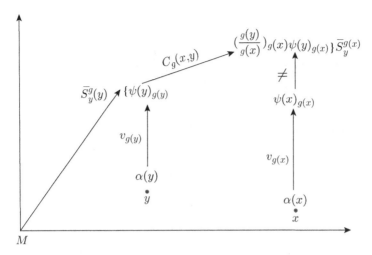

Figure 2.2. Representation of the action of the connection in mapping the number $\psi(y)_{g(y)}$ in $\bar{S}_y^{g(y)}$ to the same number $(g(y)/g(x))_{g(x)}\psi(y)_{g(x)}$ in $\bar{S}_x^{g(x)}$. The lifting or representation of the field numbers $\alpha(y)$ and $\alpha(x)$ at y and x on M to numbers in $\bar{S}_y^{g(y)}$ and $\bar{S}_x^{g(x)}$ with values $\psi(y)$ and $\psi(x)$ is shown by up pointing arrows labeled with the value functions. The line with the unequal sign indicates that the number $\psi(x)_{g(x)}$ is not related to the parallel transport of $\psi(y)_{g(y)}$ to x.

Eqs. (2.22) and (2.23). Use of these maps to rewrite Eq. (2.24) gives

$$
C_g(x,y)(\psi(y)_{g(y)}) = W_y^x Z_y^x \psi(y)_{g(y)} = W_y^x \left(\frac{g(y)_{g(y)}}{g(x)_{g(y)}} \psi(y)_{g(y)} \right)
$$

$$
= \frac{g(y)_{g(x)}}{g(x)_{g(x)}} W_y^x (\psi(y)_{g(y)}) = \left(\frac{g(y)}{g(x)} \psi(y) \right)_{g(x)} .
$$

$$(2.27)$$

The representation of the Z and W operators defined in Eqs. (2.22) and (2.23) have been changed so that the maps Z_y^x and W_y^x are number preserving value changing and number changing value preserving maps of numbers from y to x.

The connection can be used to convert ψ as a section on the bundle $\mathfrak{M}\mathfrak{F}^g$, with values in different number structures to a field with number values all in the same structure $\bar{S}_x^{g(x)}$ at a reference location, x. The result is a scaled field, ψ_x^g, where for each y in M, $\psi_x^g(y)$ is a number in $\bar{S}_x^{g(x)}$ with representation $(g(y)/g(x))_{g(x)}(\psi(y))_{g(x)}$.

That is,

$$\psi_x^g(y) = C_g(x, y)(\psi(y)_{g(y)}) = \left(\frac{g(y)}{g(x)}\right)_{g(x)} \psi(y)_{g(x)}. \qquad (2.28)$$

The values of the field, ψ_x^g, at different locations can be combined arithmetically. The reason is that the values of ψ_x^g at different points of M are all in a single structure, $\bar{S}_x^{g(x)}$. This is not the case for the field, ψ, as the numbers, $\psi(y)_{g(y)}$ and $\psi(x)_{g(x)}$, are in different complex number structures. For example, addition or subtraction of $\psi_x^g(y)$ and $\psi_x^g(z)$ in $\bar{S}_x^{g(x)}$ is given by

$$\psi_x^g(y) \pm_{g(x)} \psi_x^g(z) = C_g(x, y)(\psi(y)_{g(y)}) \pm_{g(x)} C_g(x, z)\psi(z)_{g(z)}$$

$$= \left(\frac{g(y)}{g(x)}\psi(y)\right)_{g(x)} \pm_{g(x)} \left(\frac{g(z)}{g(x)}\psi(z)\right)_{g(x)}.$$
$$(2.29)$$

The description of the connection given so far applies to the general case in which the valuation function for the field α in M is not known. In this case, $v_{g(y)}(\alpha(y)) = \psi(y)$ and $v_{g(x)}(\alpha(x)) = \psi(x)$.

The special case defines the valuation function so that $v_1(\alpha(y)) = \alpha(y)$ for all y. Number and number value are conflated for the identity scaling factor. From this restriction, the value function for value factors such as $g(y)$ is defined by

$$v_{g(y)}(\alpha(y)) = \frac{\alpha(y)}{g(y)}. \qquad (2.30)$$

Use of the connection to parallel transform the number with this value at y to x gives

$$C_g(x, y)\left[\frac{\alpha(y)}{g(y)}\right]_{g(y)} = \left[\frac{g(y)}{g(x)}\frac{\alpha(y)}{g(y)}\right]_{g(x)} = \frac{\alpha(y)}{g(x)}. \qquad (2.31)$$

For this special case, the parallel transport from y to x erases any memory of the value factor $g(y)$. The problems in combining arithmetic combinations of numbers with their associated values at different locations disappear:

$$C_g(x,y)\left(\frac{\alpha(y)}{g(y)}\right) \pm C_g(x,z)\left(\frac{\alpha(z)}{g(z)}\right) = \frac{\alpha(y) \pm \alpha(z)}{g(x)}.$$

2.3.1 *Vector fields*

The effects of the value field on vector fields are similar to the effects on scalar fields. If $\vec{\beta}$ is a vector field over M, then, for each y, $\vec{\beta}(y)$ is a vector in M. The lifting of $\vec{\beta}$ to be a section on the fiber bundle $\mathfrak{MF}^g$ gives field vectors as $\vec{\phi}(y)_{g(y)}$ as a vector in $\bar{V}_y^{g(y)}$ and $\vec{\phi}(x)_{g(x)}$ as a vector in $\bar{V}_x^{g(x)}$. The effect of the lift is represented by

$$\vec{\beta}(y) \to \vec{\phi}(y)_{g(y)} \quad \text{and} \quad \vec{\beta}(x) \to \vec{\phi}(x)_{g(x)}. \tag{2.32}$$

Here $\vec{\phi}(y)$ and $\vec{\phi}(x)$ are the values of $\vec{\beta}(y)$ and $\vec{\beta}(x)$ in their respective vector spaces, $\bar{V}_y^{g(y)}$ and $\bar{V}_x^{g(x)}$. This is represented by

$$\vec{\phi}(y) = v_{g(y)}(\vec{\beta}(y)) \quad \text{and} \quad \vec{\phi}(x) = v_{g(x)}(\vec{\beta}(x)). \tag{2.33}$$

As was the case for number fields, linear combinations of the vectors $\vec{\phi}(y)_{g(y)}$ and $\vec{\phi}(z)_{g(z)}$ are not defined. The reason is that these are vectors in different vector structures, $\bar{V}_y^{g(y)}$ and $\bar{V}_z^{g(z)}$, at different locations, y and z. Linear combinations of vectors are defined between vectors within the same structures, not between vectors in different structures.

This is fixed by the use of connections to parallel transform the vectors $\vec{\phi}(y)_{g(y)}$ for different y to the same vectors in a vector space, $\bar{V}_x^{g(x)}$, at an arbitrary reference location, x. The expression is similar to that for numbers. One has

$$\vec{\phi}_x^g(y) = C_g(x,y)\vec{\phi}(y)_{g(y)} = \left(\frac{g(y)}{g(x)}\right)_{g(x)} \vec{\phi}(y)_{g(x)}. \tag{2.34}$$

This relation holds for all y in M.

Figure 2.2, with minor changes, applies to vectors and vector values. The change consists of replacing ψ by $\vec{\phi}$ and replacing $\bar{S}_y^{g(y)}$ and $\bar{S}_x^{g(x)}$ with $\bar{V}_y^{g(y)}$ and $\bar{V}_x^{g(x)}$.

Connections are used to define linear combinations of $\vec{\phi}(y)_{g(y)}$ and $\vec{\phi}(z)_{g(z)}$ as linear combinations of $\vec{\phi}_x^g(y)$ and $\vec{\phi}_x^g(z)$. One has

$$\vec{\phi}_x^g(y) \pm_{g(x)} \vec{\phi}_x^g(z) = C_g(x,y)\vec{\phi}(y)_{g(y)} \pm_{g(x)} C_g(x,z)\vec{\phi}(z)_{g(z)}$$

$$= \left(\frac{g(y)}{g(x)}\vec{\phi}(y)\right)_{g(x)} \pm_{g(x)} \left(\frac{g(z)}{g(x)}\vec{\phi}(z)\right)_{g(x)} .$$

$$(2.35)$$

The description of the effect of the value field, g, on vectors applies to vector spaces of any dimension. For example, let the vector spaces, $\bar{V}_y^{g(y)}$, be n dimensional. A basis representation of $\vec{\phi}(y)_{g(y)}$ is

$$\vec{\phi}(y)_{g(y)} = \sum_j c_j(y)_{g(y)} |j\rangle_{g(y)}. \qquad (2.36)$$

In this equation, $c_j(y)_{g(y)}$ is a number in $\bar{C}_y^{g(y)}$ and $|j\rangle_{g(y)}$ is a basis vector in $\bar{V}_y^{g(y)}$.

Equation (2.34) is used to define $[\vec{\phi}^g(y)]_x$, from $\vec{\phi}(y)_{g(y)}$ as represented by Eq. (2.36). For each y, the vector, $\vec{\phi}_x^g(y)$, is an element of $\bar{V}_x^{g(x)}$. One has

$$\vec{\phi}_x^g(y) = \left[\frac{g(y)}{g(x)}\sum_j c_j(y)|j\rangle\right]_{g(x)} . \qquad (2.37)$$

One feature worthy of note is that parallel transformation of the norm of $\vec{\phi}(y)_{g(y)}$ is the same number as is the norm of the transformed vector $\vec{\phi}_x^g(y)$. This follows from

$$C_g(x,y)(|\vec{\phi}(y)|)_{g(y)} = \left(\frac{g(y)}{g(x)}|\vec{\phi}(y)|\right)_{g(x)}$$

$$= \left(\left|\frac{g(y)}{g(x)}\vec{\phi}(y)\right|\right)_{g(x)} = (|\vec{\phi}_x^g(y)|)_{g(x)}.$$

$$(2.38)$$

This result is valid if and only if the values of g are all positive real numbers. This is the case here.

As was the case for numbers, much of the description for vectors is based on a general value function. No restrictions were placed on the dependence of the value function on the value factor, $g(x)$.

If the same restriction on the value function for numbers holds for vectors, then for each vector $\vec{\beta}$ in M $v_1(\vec{\beta}) = \vec{\beta}$. Use of the connection for vectors gives

$$v_{g(y)}(\vec{\beta}) = \frac{\vec{\beta}}{g(y)}. \tag{2.39}$$

This result normalizes the length of the vector to the norm $|\vec{\beta}|$ at locations y, where $g(y) = 1$.

2.4 Integrals and derivatives of bundle sections

One consequence of the mathematical arena, based on local mathematics and the presence of the value field, is that fields, as functions from M to either numbers or vectors, are treated as sections on a fiber bundle. Here the fields are treated as sections on the bundle $\mathfrak{M}\mathfrak{F}^g$. One consequence of this is that integrals and derivatives of fields, as bundle sections, are not defined. The reason is that both integrals and derivatives are defined by arithmetic combinations of numbers at different points of M. Arithmetic combinations of numbers or vectors in structures at different points are not defined. These are defined only within number or vector structures at the same point.

This is remedied by the use of connections to parallel transport section numbers or vectors to a common point. Integration or derivation is defined because the implied arithmetic combination of numbers or vectors occurs within the same structure. The descriptions for integrals and derivatives are given in the following sections.

2.4.1 *Integrals*

Let α be a number field on M and $\psi(-)_{g(-)} = [v_{g(-)}(\alpha(-))]_{g(-)}$ be the lift of α as a number section on $\mathfrak{M}\mathfrak{F}^g$. The integral

$$I(\psi_g) = \int [v_{g(y)}(\alpha(y))]_{g(y)} dy_{g(y)} = \int \psi(y)_{g(y)} dy_{g(y)} \tag{2.40}$$

over M of the section is not defined. For each y, the integrand $\psi(y)_{g(y)}$ is a number in $\bar{S}_y^{g(y)}$. The implied summation over y in the definition of the integral is not defined as it is between numbers in structures at different points so M.

This is fixed by the use of connections to parallel transport the numbers, $\psi(y)_{g(y)}$, to numbers in a single complex number structure at an arbitrary reference point. If x is the reference point, then the transported section numbers

$$C_g(x,y)\psi(y)_{g(y)} = \left[\frac{g(y)}{g(x)}\right]_{g(x)} \psi(y)_{g(x)} = \left[\frac{g(y)}{g(x)}\psi(y)\right]_{g(x)} \tag{2.41}$$

are all numbers in $\bar{S}_x^{g(x)}$. As a result, the section integral

$$[I(\psi)]_{g(x)} = \left[\frac{1}{g(x)}\right]_{g(x)} \int [g(y)\psi(y))]_{g(x)} \, dy_{g(x)} \tag{2.42}$$

is well-defined. The factor, $(1/g(x))_{g(x)}$, has been brought outside the integral as it is independent of the integration variable.

Equations (2.41) and (2.42) are valid in the case that M is multidimensional. As an example if M is three dimensional, then $y = y_1, y_2, y_3$. The integral becomes a multidimensional integral as in

$$I(\psi)_{g(x)} = \left[\frac{1}{g(x)} \int \int \int g(y)\psi(y)dy_1 dy_2 dy_3\right]_{g(x)}. \tag{2.43}$$

Here $x = x_1, x_2, x_3$.

Similar considerations hold for integrals of a vector section $\vec{\phi}(-)_{g(-)}$, on M. Use of the connection, $C_g(x,y)$, to parallel transform the vector integrand to a reference point, x, gives

$$I(\vec{\phi})_{g(x)} = \int \left[\frac{g(y)}{g(x)}\vec{\phi}(y)dy\right]_{g(x)} = \left[\frac{1}{g(x)}\int g(y)\vec{\phi}(y)dy\right]_{g(x)}. \tag{2.44}$$

For each y, the integrands are all vectors in the same vector structure, $\vec{V}_x^{g(x)}$.

For the restricted valuation case where $v_1(\alpha(y)) = \alpha(y)$,

$$\psi(y)_{g(y)} = \left[\frac{\alpha(y)}{g(y)}\right]_{g(y)}. \tag{2.45}$$

Use of $C_g(x,y)$ to parallel transport $\psi(y)_{g(y)}$ to x gives

$$C_g(x,y)\left[\frac{\alpha(y)}{g(y)}\right]_{g(y)} = \left[\frac{\alpha(y)}{g(x)}\right]_{g(x)}. \tag{2.46}$$

Use of this in the expression for the integral gives

$$I(\alpha)_{g(x)} = \int \left[\frac{1}{g(x)}\alpha(y)dy\right]_{g(x)} = \left[\frac{1}{g(x)}\int \alpha(y)dy\right]_{g(x)}. \tag{2.47}$$

This shows that the value of the integral depends on the location in M.

Application of this representation of integrals to number or vector fields as sections and to physics and geometry requires extension to sections that depend on more than one point of M. The simplest case is that in which a field $\theta(y,z) = \psi(y)\gamma(z)$ is a scalar product of two number fields, ψ and γ. The usual fiber bundle, $\mathfrak{MF}^g$, can be used because both ψ and γ become sections on the same bundle.

In this framework, $\theta(y,z)$ becomes

$$\theta(y,z)_{g(y),g(z)} = \psi(y)_{g(y)}\gamma(z)_{g(z)}. \tag{2.48}$$

This scalar product is not defined because the factors are in different complex number structures.

There are three ways to fix this. Parallel transport of $\psi(y)_{g(y)}$ to z, transport of $\gamma(z)_{g(z)}$ to y, or transport of both factors to a common point, x. The results of these transports are

$$\theta_1(y,z)_{g(z)} = (C_g(z,y)\psi(y)_{g(y)})\gamma(z)_{g(z)} = \left[\frac{g(y)}{g(z)}\psi(y)\gamma(z)\right]_{g(z)}, \tag{2.49}$$

$$\theta_2(y,z)_{g(y)} = \psi(y)_{g(y)}(C_g(y,z)\gamma(z)_{g(z)}) = \left[\frac{g(z)}{g(y)}\psi(y)\gamma(z)\right]_{g(y)}, \tag{2.50}$$

and

$$\theta_3(y,z)_{g(x)} = (C_g(x,y)\psi(y)_{g(y)})C_g(x,z)\gamma(z)_{g(z)}$$

$$= \left[\frac{g(y)}{g(x)}\frac{g(z)}{g(x)}\psi(y)\gamma(z)\right]_{g(x)}$$

$$= \left[\frac{1}{g(x)^2}g(y)\psi(y)g(z)\gamma(z)\right]_{g(x)}. \qquad (2.51)$$

The y integral of $\theta_1(y,z)$ is defined but the z integral is not defined. The opposite is true for the integral of $\theta_2(y,z)$. These problems can be fixed by defining $\theta_1(x,z)$ and $\theta_2(x,y)$ to be the respective parallel transports of z to x and y to x. One obtains

$$\theta_1(y,x)_{g(x)} = C_g(x,z)(C_g(z,y)(\psi(y)_{g(y)})\gamma(z)_{g(z)})$$

$$= C_g(x,z)\left[\frac{g(y)}{g(z)}\psi(y)\gamma(z)\right]_{g(z)} = \left[\frac{g(y)}{g(x)}\psi(y)\gamma(z)\right]_{g(x)}$$

$$(2.52)$$

and

$$\theta_2(z,x) = C_g(x,y)(\psi(y)_{g(y)}(C_g(y,z)\gamma(z)_{g(z)}))$$

$$= C_g(x,y)\left[\frac{g(z)}{g(y)}\psi(y)\gamma(z)\right]_{g(y)} = \left[\frac{g(z)}{g(x)}\psi(y)\gamma(z)\right]_{g(x)}.$$

$$(2.53)$$

The three integrals become

$$\left[\int \theta_1(y,z)dydz\right]_{g(x)} = \left[\frac{1}{g(x)}\int g(y)\psi(y)\gamma(z)dydz\right]_{g(x)}, \qquad (2.54)$$

$$\left[\int \theta_2(y,z)dydz\right]_{g(x)} = \left[\frac{1}{g(x)}\int g(z)\psi(y)\gamma(z)dydz\right]_{g(x)}, \qquad (2.55)$$

and

$$\left[\int \theta_3(y,z)dydz\right]_{g(x)} = \left[\frac{1}{g(x)^2}\int g(y)g(z)\psi(y)\gamma(z)dydz\right]_{g(x)}. $$

$$(2.56)$$

The difference in values of these three integrals shows the effect of variations in the value function, g, on space or space–time. The first two integrals are asymmetric in that the effect of the value function shows up on just one of the component factors, θ or γ. They have the advantage that transport of the integrals to another reference location, v, replaces $g(x)$ with $g(v)$. There is no memory of first going to x.

The third integral is symmetric in that the effect of g applies to both components, θ and γ. Parallel transport of the integral to v replaces $g(x)^2$ with $g(x)g(v)$. This shows that it has the disadvantage of retaining a memory of the first transport of x. This is problematic because transport of mathematical elements from one location to another followed by transport to a third location should retain no memory of transport to the middle location. This is a consequence of the transitivity of the connection in that

$$C_g(x, y)C_g(y, z) = C_g(x, z). \tag{2.57}$$

This problem will be taken up again later on.

The use of connections to define integrals of $\theta(y, z)$ where $\theta(y, z)$ is not a product of two functions requires a change in the definition of connections. The fiber bundle is almost the same as that already introduced. The difference between the fiber bundle used here and that already described accounts for the need to lift the values of $\theta(y, z)$ to number structures associated with each pair, y, z, of locations on M.

Let

$$\mathfrak{MF}^h = M \times F, \pi_h, M \tag{2.58}$$

be a fiber bundle. The fiber has the same definition as for the bundle with g the scaling factor. It is given by

$$F = \cup_r \cup_S S^r. \tag{2.59}$$

The union over S is over all types of mathematical systems that include scalars in their description. The union over r is over all real scaling factors.

The difference between the bundles, $\mathfrak{M}\mathfrak{F}^g$ and $\mathfrak{M}\mathfrak{F}^h$, is in the definition of h. Here h is a real value function that assigns a real number value to each pair of locations in M. The inverse of the projection operator reflects this difference in that

$$\pi_h^{-1}(y, z) = F_{y,z}^{h(y,z)}. \tag{2.60}$$

Here $F_{y,z}^{h(y,z)}$ is the fiber associated with the location pair, y, z, in M.

Let β be a complex valued function defined on pairs of locations in M. For each y, z, the lift of the number $\beta(y, z)$ to the fiber bundle gives

$$[v_{h(y,z)}(\beta(y, z))]_{h(y,z)} = \theta(y, z)_{h(y,z)} \tag{2.61}$$

as a number in $\bar{C}_{y,z}^{h(y,z)}$. As was the case for the product of two component functions, the integral

$$\int \theta(y, z)_{h(y,z)} dy_{h(y,z)} dz_{h(y,z)} \tag{2.62}$$

is not defined. The problem is that the implied summation in the definition of the integral is not defined between structures at different pairs of locations.

Parallel transport of the integrand to a reference location pair, x, v, fixes this problem. Transport is defined by the connection $C_h(v, x; y, z)$. Parallel transport of $\theta(y, z)_{h(y,z)}$ to the reference location pair, v, x, is defined by

$$\theta(y, z)_{h(y,z)} \to C_h(v, x; y, z)\theta(y, z)_{h(y,z)} = \left[\frac{h(y, z)}{h(v, x)}\theta(y, z)\right]_{h(v,x)}. \tag{2.63}$$

The integral of Eq. (2.62) becomes

$$I(\theta)_{h(v,x)} = \left[\frac{1}{h(v, x)}\right]_{h(v,x)} \int [h(y, z)\theta(y, z) dy dz][h(v, x)]. \tag{2.64}$$

The factor $1/h(v, x)$ has been moved outside the integral as it is independent of the integrand variables.

The h value function can be defined from the g value function. A reasonable definition of the h function is as the square root of the product of the g function for pairs of locations. For each pair y, z of locations, $h(y, z)$ is defined by

$$h(y, z) = (g(y)g(z))^{1/2}. \tag{2.65}$$

Use of this in Eq. (2.63) gives

$$C_h(v, x; y, z)\theta(y, z)_{h(y,z)} = \left[\frac{(g(y)g(z))^{1/2}}{(g(v)g(x))^{1/2}} \theta(y, z) \right]_{(g(v)g(x))^{1/2})}. \tag{2.66}$$

A good feature of this definition of h is that if $v = x$, then $h(x, x) = g(x)$. This shows that the reference structures at single locations can be used for functions $\theta(y)$ and $\theta(y, z)$. The definition is symmetric in the effect of scaling in that the g value functions are present for both y and z. The definition is also transitive in that transport of $I(\theta)_{(g(v)g(x))^{1/2}}$ to another location pair, w, u, gives the same integral as does use of w, u as the initial reference location pair. This is a consequence of the transitivity of the connection $C_h(v, x; y, z)$.

The problems in use of parallel transport to obtain an integral of $\theta(y, z) = \phi(y)\gamma(z)$ as a product, shown in Eqs. (2.54)–(2.56), can be avoided by use of the connection, $C_h(x, x; y, z)$. Lift of the product $\psi(y)\gamma(z)$ to numbers in the local complex number structures, $\bar{C}_y^{g(y)}$ and $\bar{C}_z^{g(z)}$, gives the product, $\psi(y)_{g(y)}\gamma(z)_{g(z)}$. As noted this product is not defined. Parallel transport of this product to x gives a defined integrand,

$$[\theta(y, z)]_{h(y,z)} \to C_h(x, x; z, y)\psi(y)_{g(y)}\gamma(z)_{g(z)}$$
$$= \left[\frac{1}{g(x)} \sqrt{g(y)g(z)}\psi(y)\gamma(z) \right]_{g(x)}. \tag{2.67}$$

The integral of this integrand is defined because the integrands are all numbers in $\bar{C}_x^{g(x)}$.

This recipe for defining integrals of functions on two variables extends to functions of n variables, as in $\theta(y_1, \ldots, y_n)$. The value function, $h(y, z)$, becomes $h(y_1, \ldots, y_n)$, where

$$h(y_1, \ldots, y_n) = (g(y_1)g(y_2) \cdots g(y_n))^{1/n}. \qquad (2.68)$$

The connection $C_h(x; x_1, \ldots, x_n)$ parallel transports the number, $[\theta(y_1, \ldots, y_n)]_{h(y_1, \ldots, y_n)}$ in $\bar{C}_{y_1, \ldots, y_n}^{h(y_1, \ldots, y_n)}$ to a number in $\bar{C}_x^{g(x)}$. The value change factor is given by $h(y_1, \ldots, y_n)/g(x)$.

For the special case where $v_1(\beta(y, z)) = \beta(y, z)$, one has

$$v_{h(y,z)}(\beta(y, z)) = \frac{1}{h(y, z)}\beta(y, z).$$

The integral of Eq. (2.64) becomes

$$I(\beta)_{h(v,x)}) = \left[\frac{1}{h(v, x)} \int \beta(y, z)dydz \right]_{h(v,x)}. \qquad (2.69)$$

This integral depends on the reference location pair, v, x. The dependence on $h(y, z)$ disappears.

It should be emphasized that the change in the reference locations is a translation of the mathematical description of the properties of the scalar and vector fields from structures associated with a set of locations to mathematical structures associated with a different location set. It does not correspond to a translation of the fields themselves. Their location and distribution on M stay the same.

2.4.2 *Derivatives*

Connections also play a role in derivatives over M of scalar or vector fields as sections on the fiber bundle. Since this is well known and used in gauge theories [16,45], the description here will be relatively brief. A description for the effect of a general valuation function is followed by a description for the specific function where number and vector are conflated with number value and vector value for the identity value factor.

With fiber bundles and local mathematics as a background, let $\psi(-)_{g(-)}$ be a scalar section on the bundle, $\mathfrak{M}\mathfrak{F}^g$. For each point y

in M, $\psi(y)_{g(y)} = [v_{g(y)}(\alpha(y))]_{g(y)}$ is a number in $\bar{S}_y^{g(y)}$ with number value $\psi(y)$. As before, α is a field of numbers on M.

Let M be n dimensional. The partial derivative in the direction $\hat{j}$ with $j = 1, \ldots, n$ is given by

$$\frac{\partial}{\partial^j y}(\psi(y)_{g(y)}) = \lim_{dy_j \to 0} \frac{\psi(y + dy_j)_{g(y+dy_j)} - \psi(y)_{g(y)}}{dy_j}. \qquad (2.70)$$

Here

$$\psi(y + dy_j)_{g(y+dy_j)} = v_{g(y+dy_j)}(\alpha(y + dy_j)). \qquad (2.71)$$

As might be guessed, the partial derivative is not defined. The subtraction shown in the numerator is not defined as the two components are in different complex number structures, $\bar{S}_{y+dy_j}^{g(y+dy_j)}$ and $\bar{S}_y^{g(y)}$. Parallel transport of the left hand numerator component to location y gives

$$[D_{g,j,y}\psi(y)]_{g(y)} = \frac{1}{(dy_j)_{g(y)}}[C_g(y, y + dy_j)\psi(y + dy_j) - \psi(y)]_{g(y)}$$

$$= \frac{1}{(dy_j)_{g(y)}}\left[\frac{g(y + dy_j)}{g(y)}\psi(y + dy_j) - \psi(y)\right]_{g(y)}. \qquad (2.72)$$

The limit, $dy_j \to 0$, is understood.

Taylor expansion to first order of $g(y + dy_j)$ gives

$$g(y + dy_j) \approx g(y) + dy_j \frac{\partial g(y)}{\partial^j y}. \qquad (2.73)$$

Insertion of this into Eq. (2.72) gives

$$[D_{g,j,y}\psi(y)]_{g(y)} = \left[\left(\frac{\partial}{\partial^j y} + \frac{1}{g(y)}\frac{\partial g(y)}{\partial^j y}\right)\psi(y)\right]_{g(y)}. \qquad (2.74)$$

This result shows that the presence of the g field and local mathematics introduces an extra term, $\partial_{j,y}g(y)/g(y) = (1/g(y))(\partial g(y)/\partial^j y)$, into the expression for the derivative of ψ as a section on the fiber bundle.

The same prescription applies to vector fields $\vec{\phi}$. If $\vec{\phi}$ is a section on the vector spaces in the fiber bundle, $\mathfrak{M}\mathfrak{F}^g$, Eq. (2.12), then the derivative, given by Eq. (2.70) with $\vec{\phi}$ replacing ψ, is not defined. The vectors $\vec{\phi}(y+d^j y)_{g(y+d^j y)}$ and $\vec{\phi}(y)_{g(y)}$ belong to different vector spaces, $\bar{V}_{y+d^j y}^{g(y+d^j y)}$ and $\bar{V}_y^{g(y)}$, and subtraction is not defined between the two spaces.

This problem is fixed by following the same steps outlined above for scalar fields. The resulting derivative is given by

$$[D_{g,j,y}\vec{\phi}(y)]_{g(y)} = \left[\partial_{j,y} + \left(\frac{\partial_{j,y}g(y)}{g(y)}\right)\vec{\phi}(y)\right]_{g(y)}. \qquad (2.75)$$

This equation gives the derivative for vectors in the presence of the g field. The equivalent expression for vector values is obtained by removal of the subscript, $g(y)$, on both sides of the equation.

For the case where the value function satisfies the restriction that $v_1(\alpha(y)) = \alpha(y)$, one knows that $v_{g(y)}(\alpha(y)) = \alpha(y)/g(y)$. The relation between α and the value field ψ gives $\psi(y)_{g(y)} = [\alpha(y)/g(y)]_{g(y)}$ and

$$\psi(y + dy_j)_{g(y+dy_j)} = \left[\frac{\alpha(y + dy_j)}{y(y + dy_j)}\right]_{g(y+dy_j)}.$$

Parallel transport to location y gives

$$C_g(y, y + dy_j)\left[\frac{\alpha(y + dy_j)}{g(y + dy_j)}\right]_{g(y+dy_j)}$$

$$= \left[\frac{g(y + dy_j)}{g(y)}\frac{\alpha(y + dy_j)}{g(y + dy_j)}\right]_{g(y)} = \left[\frac{1}{g(y)}\alpha(y + dy_j)\right]_{g(y)}. \qquad (2.76)$$

Use of this in the expression for the derivative in Eq. (2.72) gives the usual expression for the derivative

$$[D_{g,j,y}\psi(y)]_{g(y)} = \left[\frac{\partial_{y,j}\alpha(y)}{g(y)}\right]_{g(y)}. \qquad (2.77)$$

The extra term in $D_{g,j,y}$ in Eq. (2.74) is missing. The y location dependence of the derivative remains.

The same result holds for a vector field. If $\vec{\beta}$ is a vector field on M, then $v_{g(y)}(\vec{\beta}) = \vec{\beta}/g(y)$. Proceeding through the same steps as were done for the scalar field gives the derivative expression

$$[D_{g,j,y}\vec{\phi}(y)]_{g(y)} = \left[\frac{\partial_{y,j}\vec{\beta}(y)}{g(y)}\right]_{g(y)}. \tag{2.78}$$

Here $\vec{\phi}(y) = \vec{\beta}(y)/g(y)$.

In Part II, the material developed in this and earlier chapters will be used to show the effect of local mathematics and the presence of the g field on physical and geometric quantities. For this, the base manifold, M, will be taken to be either three-dimensional Euclidean space, Euclidean space and time, or four-dimensional space–time as Minkowski space. In addition, an alternative exponential expression of the g field as

$$g(y) = e^{\alpha(y)} \tag{2.79}$$

will be used. This representation with $-\infty < \alpha(y) < \infty$ is closer to the representation of gauge and other fields in physics than is the more compact notation used in Part I. It also limits $g(y)$ to be a positive real number value.

Part II
Effects of Local Mathematics and the Value Field on Physical and Geometric Quantities

Chapter 3

Introduction

The mathematical material presented in the first part forms the arena for a description of the effect of local mathematics and the presence of the number value field g (or α, Eq. (2.79)) on theoretical descriptions of physical and geometric quantities. As noted in the introduction, the existence of the α field is a consequence of the freedom of choice or "no information at a distance" principle. Local mathematical structures, such as different types of numbers, vector spaces, and other types of structures, and the presence of the α field, are the bases for the use of fiber bundles as the mathematical arena for a theoretical description of many physical and geometric quantities.

Physical and geometric quantities expressed as integrals or derivatives over space, time, or space–time begin with the integrand or field as a section on the fiber bundle. Since the section components for different locations belong to separate mathematical structures, the integrals or derivatives of the sections are not defined. This is a consequence of the fact that the arithmetic combinations of the section numbers or vectors at different locations, needed in the definition of integrals or derivatives, are not defined.

The use of connections to solve these problems is shown in Sections 2.4.1 and 2.4.2. The connections parallel transport section components to a single number or vector structure at a common reference location. The price to be paid for having defined integrals or derivatives is the introduction of location-dependent α-based factors multiplying the integrands or field component numbers or vectors.

As noted in the above, the value field will be described with the exponential α field instead of the g field. The relation between the two is given in Eq. (2.79). The α representation is more suited to the description of fields and quantities that depend on space or space–time locations. The notationally simpler g representation will be used as subscript locations for section components and value factors for structures.

In reading this part of the book, it is important to keep some basic aspects of this work in mind. Theoretical physics, as descriptions or predictions of properties of physical systems, have meaning or value. All mathematical expressions, as equations or in any other form, used in physics have value or meaning. Numbers and vectors appear in theory as number values and vector values. Operations on numbers and vectors have meaning.

By themselves, numbers, vectors, and other mathematical elements have no intrinsic meaning. This is especially evident in their representations by symbols or symbol strings. The form of the representations allows one to determine whether they are numbers of some type or vectors. Their meanings are not known until the structures containing them are specified.

In this part of the book, the effect of the value or meaning field α on theory in different areas of physics and geometry will be described. An essential part of the effect is based on the relation between numbers and their values at different locations. This is made explicit by letting β be a number at location y in M. The general relation between β and its value at y is given by $v_{g(y)}(\beta)$.

In Part II, the value relation will be restricted to be that shown in Eq. (1.4). The value function is required to satisfy the following conditions: If β is a number at location y, then

$$v_1(\beta) = \beta \quad \text{and} \quad v_{g(y)}(\beta) = \frac{\beta}{g(y)}. \tag{3.1}$$

This equation shows that if $g(y) = 1$, then β is identified with its value. The two concepts number and number value are conflated. If $g(y) \neq 1$, then the value of the number β is $\beta/g(y)$. Number and number value are distinct.

The effect of this restricted value function will be examined on theoretical description of quantities in different areas of physics and geometry. For quantities expressed by integrals or derivatives of functions or fields of space, time, or space–time, the effect of the restricted value function is the introduction of a space, time, or space–time location dependence of the quantity, Sections 2.4.1 and 2.4.2. There is no effect on the value of the integral or derivative itself.

This effect is shown in the following few chapters for fields in gauge theories, for properties of states in quantum mechanics, and on spaces at each space–time point in the universe. For all these quantities, the theory descriptions include an overall location-dependent scale factor. It follows that their meaning or value is different at different locations. If $v_{g(y)}(Q)$ is the value or meaning of a theory description of a physical quantity, then $v_{g(y)}(Q) = Q/g(y)$ is the meaning of Q at y. Note that the usual location independent meaning of Q corresponds to the special case where $g(y) = 1$ for all y.

So far, to date, no local experiment has been interpreted to show the effect of the presence of the α field. It follows that any local space or time variation of the field must be too small to be detected experimentally. For many types of physical systems, theoretical descriptions formulated with and without the value field are experimentally indistinguishable.

The lack of local experimental support for the presence of the value field for many types of physical systems does not mean that it has no effect on physics. It does mean that in local regions of the universe that are occupiable by us as observers, the space and time dependence of the α field is too small to be observed experimentally.

Chapter 4

Gauge Theories

4.1 Introduction

Local mathematics and the presence of the field α are well suited to describe an extension of gauge theories to include the effect of the α field. The reason is that the mathematical background of gauge theories already includes separate vector spaces at each point of space–time, M. Matter fields as fermion fields, ψ, take values in the different spaces in that $\psi(y)$ is a vector in a vector space at y. The freedom of basis choice in the different vector spaces is accounted for by unitary operators, $U_P(x, y)$, that map vectors in a vector space at y along a path P to those in a space at x. These operators are elements of a gauge group. Physics makes use of the groups, $U(1), SU(2)$, and $SU(3)$. The gauge group for the standard model, which underlies much of physics, is the product of these three groups.

The extension of gauge theories is based on the fact that vector spaces include scalars as part of their axiomatic properties. Therefore, it seems reasonable to extend the mathematical arena of gauge theories to include local scalar structures with the vector spaces. For gauge theories, the scalars are the complex numbers.

Fiber bundles provide a good way to describe the extensions of gauge theories to include the effects of the real scalar field, g.

The fiber bundle, $\mathfrak{M}\mathfrak{F}^g$, of Eq. (2.12) is useful for this purpose. For gauge theories, the bundle fiber at y is[1] $F_y^{g(y)} = \bar{C}_y^{g(y)} \times \bar{V}_y^{g(y)}$.

One begins with fields for different physical particles as scalars or vectors on M. The lift of fields to be sections on the fiber bundle gives fields whose elements acquire meaning from the structures colocated with the field elements. If the lifted field, $\psi(-)_{g(-)}$, is a complex scalar field, then for each y, $\psi(y)_{g(y)}$ is a number in $\bar{C}_y^{g(y)}$ with number value $\psi(y)$. If $\psi(-)_{g(-)}$ is a vector field, then $\psi(y)_{g(y)}$ is a vector in $\bar{V}_y^{g(y)}$ with vector value $\psi(y)$.

In gauge theories, the definition of covariant derivatives requires the introduction of gauge groups. The extension used here includes the effect of the scalar g field. This is implemented by a connection that includes an extra scalar term. For Abelian gauge theories, there is an additional component of the connection arising from the action of a unitary operator, $U(x, y)$ [46], in the group $U(1)$. For non-Abelian theories, $U(x, y)$ is a unitary operator in the group $SU(2)$ or $SU(3)$ (for the standard model).

The simplest case to consider is that for $\psi(-)_{g(-)}$ as a complex scalar field. A description of the effect of the α field on the Klein–Gordon (KG) Lagrangian occupies the following section. This is followed by a section on the effect of α on the Dirac Lagrangian and the equation for ψ as a vector field. Next come sections on extensions to Abelian and non-Abelian theories. Recall that $g(y) = e^{\alpha(y)}$.

4.2   Klein–Gordon fields

The KG Lagrangian density at location y in space–time for a scalar field is given by

$$\mathcal{L}_{KG}(\psi(y)) = |\partial_{\mu,y}\psi|^2 - m^2|\psi(y)|^2 \tag{4.1}$$

where

$$|\partial_{\mu,y}\psi|^2 = \eta^{\mu\nu}\frac{\partial\psi(y)^*}{\partial y^\mu}\frac{\partial\psi(y)}{\partial y^\nu}. \tag{4.2}$$

---

[1]The g field will continue to be used as superscripts and subscripts to denote value factors for scalars and vectors.

Sum over repeated indices is implied and $y = t, \mathbf{y}$. The metric tensor, η, is given by $\text{diag}(1, -1, -1, -1)$.

The KG Lagrangian density for scalar fields of the form $\psi(-)_{g(-)}$ is not valid. The reason is that the derivative term,

$$[|\partial_{\mu,y}\psi|^2]_{g(y)}$$

in Eq. (4.1) is not defined.

This problem is fixed by the use of Eqs. (2.72)–(2.74). The derivative term becomes

$$[|D_{\mu,y}\psi|^2]_{g(y)}$$

where

$$D_{\mu,y} = \partial_{\mu,y} + A_\mu(y). \tag{4.3}$$

This equation for $D_{\mu,y}$ is obtained by replacing $g(y)$ by its exponential equivalent, $e^{\alpha(y)}$, in Eq. (2.74). The factor $\partial_{\mu,y}(g(y))/g(y)$ in Eq. (2.74) becomes $A_\mu(y)$, where $\vec{A}$ is the gradient field of α.

Use of this result in the Lagrangian of Eq. (4.1) gives

$$\begin{aligned}
[\mathcal{L}_{KG}(\psi(y))]_{g(y)} &= \left[|D_{\mu,y}\psi|^2 - m^2|\psi(y)|^2\right]_{g(y)} \\
&= \left[|\partial_{\mu,y}\psi + aA_\mu(y)\psi(y)|^2 - m^2|\psi(y)|^2\right]_{g(y)} \\
&= \left[|\partial_{\mu,y}\psi|^2 + (a^2 A^\mu(y)A_\mu(y) - m^2)|\psi(y)|^2\right. \\
&\quad \left. + aA^\mu(y)\left(\psi(y)\partial_{\mu,y}\psi^* + \psi(y)^*\partial_{\mu,y}\psi\right)\right]_{g(y)}.
\end{aligned} \tag{4.4}$$

A coupling constant, a, expressing the strength of the coupling between $\vec{A}$ and ψ has been added.

In the above and from now on, the metric tensor is implied in the sums over μ. Repetition of indices, such as μ, implies sums over the indices.

This equation shows the presence of the α field adds two interaction terms, $aA^\mu(y)\left(\psi(y)\partial_{\mu,y}\psi^* + \psi(y)^*\partial_{\mu,y}\psi\right)$, to the Lagrangian. These describe the interaction between ψ and $\vec{A}$. A shift of the mass squared of ψ from m^2 to $a^2 A^\mu(y)A_\mu(y) - m^2$ is also present. The Lagrangian shows that the mass shift component is space–time dependent in that the shift part, $A^\mu(y)A_\mu(y)$, depends on the location y.

The KG equation of motion is obtained by minimization of the action for the Lagrangian. The action integral is

$$S_{KG} = \int \mathcal{L}_{KG}(\psi(y))dy. \tag{4.5}$$

The integral is four-dimensional.

Here, the problem of undefined integrals, discussed in Section 2.4.1, applies to the action integral. Since the field, $\psi(-)_{g(-)}$, is a section on the fiber bundle, so is the Lagrangian density a section on the fiber bundle, $\mathfrak{M}\mathfrak{F}^g$. For each y, $\mathcal{L}_{KG}(\psi(y))$ is a real-valued physical quantity in $\bar{R}_y^{g(y)}$ in the fiber at y.[2]

From Section 2.4.1, one sees that the undefinability problem is fixed by parallel transport of the integrand to a reference point, x. The transported action is given by

$$\begin{aligned}
[S_{KG}]_{g(x)} &= \int C_g(x,y)[\mathcal{L}_{KG}(\psi(y))dy]_{g(y)} \\
&= \left[e^{-\alpha(x)} \int e^{\alpha(y)}\mathcal{L}_{KG}(\psi(y))dy \right]_{g(x)} \tag{4.6} \\
&= \left[e^{-\alpha(x)} \int e^{\alpha(y)}\left(|D_{\mu,y}\psi|^2 - m^2|\psi(y)|^2\right)dy \right]_{g(x)}.
\end{aligned}$$

The presence of the exponential factor in the equation for the action adds an extra term to the KG equation of motion. This can be seen from the derivation of the Euler–Lagrange equations for the action minimization. Replacement of S by $S+\delta S$ and ψ^* by $\psi^*+\delta\psi^*$ in the integral for the action and removal of the integral component

---

[2]The contents of the fiber of the bundle will be changed as needed for different problems. This includes adding structures for different types of numbers, vector spaces with a different number of dimensions, and other mathematical elements.

for S gives

$$\delta S = \int e^{\alpha(y)} \left(D_y^\mu \delta\psi^* D_{\mu,y}\psi - m^2 \delta\psi^*\psi \right) dy$$

$$= \int \left(e^{\alpha(y)} \partial_y^\mu (\delta\psi^*) D_{\mu,y}\psi + e^{\alpha(y)} \left(A^\mu(y) D_{\mu,y}\psi - m^2\psi \right) \delta\psi^* \right) dy$$

$$= 0. \tag{4.7}$$

Integration of the left-hand term of the integrand by parts and setting $\delta\psi^* = 0$ at the integral endpoints gives

$$\int \left(-\partial_y^\mu (e^{\alpha(y)} D_{\mu,y}\psi) + e^{\alpha(y)} \left(A^\mu(y) D_{\mu,y}\psi - m^2\psi \right) \right) \delta\psi^* dy = 0. \tag{4.8}$$

Evaluation of the derivative gives

$$\int e^{\alpha(y)} \left(\left(A^\mu(y) D_{\mu,y}\psi - m^2\psi \right) - \left(A^\mu(y) + \partial_y^\mu \right) D_{\mu,y}\psi \right) \delta\psi^* dy = 0. \tag{4.9}$$

Since neither $e^{\alpha(y)}$ nor $\delta\psi^*$ are 0, one obtains

$$-\left(A^\mu(y) + \partial_y^\mu \right) D_{\mu,y}\psi + A^\mu(y) D_{\mu,y}\psi - m^2\psi = 0 \tag{4.10}$$

as the KG equation of motion. The coupling constant, a, has been suppressed in the derivation.

This derivation can be considered to use numbers and vector components, with suppression of the $g(x)$ subscript. It can also be considered to be done in terms of values of numbers and vector components. The factor, $e^{-\alpha(x)}$, has been suppressed as it does not contribute to the equation of motion.

Eq. (4.10) can also be expressed by

$$D_y^\mu D_{\mu,y}\psi - A^\mu(y) D_{\mu,y}\psi + m^2\psi = 0 \tag{4.11}$$

or by

$$\partial_y^\mu D_{\mu,y}\psi + m^2\psi = 0. \tag{4.12}$$

This equation is obtained by the cancellation of the two terms, $A^\mu(y) D_{\mu,y}\psi$, each with opposite signs, in Eq. (4.10). Replacement of $D_{\mu,y}$ by its components and expansion of the derivative gives

$$\partial_y^\mu D_{\mu,y}\psi + m^2\psi = \partial_y^\mu \partial_{\mu,y}\psi + (\partial_y^\mu A_\mu(y))\psi + A_\mu(y)\partial_y^\mu\psi + m^2\psi = 0. \tag{4.13}$$

The presence of the two terms with the $\vec{A}$ field component and the presence of a term with the first-order derivative of ψ make this equation more difficult to solve than the usual one, where $\vec{A} = 0$ everywhere.

4.3 Dirac fields

The Dirac Lagrangian density in the presence of the α field is given by

$$
\begin{aligned}
\mathcal{LD}(y) &= \bar{\psi}(y)(i\gamma^\mu D_{\mu,y}\psi) - m\bar{\psi}(y)\psi(y) \\
&= \bar{\psi}(y)i\gamma^\mu(\partial_{\mu,y} + cA_\mu(y))\psi - m\bar{\psi}(y)\psi(y).
\end{aligned}
\tag{4.14}
$$

A coupling constant, c, has been added to give the strength of the interaction of the $\vec{A}$ field with ψ. As was the case for the KG Lagrangian density, the representation of the Dirac field as a section on the fiber bundle means that the usual partial derivatives are not defined. This is fixed by the replacement of the partial derivatives by $D_{\mu,y} = \partial_{\mu,y} + A_\mu(y)$ and $D_y^\mu = \partial_y^\mu + A^\mu(y)$.

The Dirac equation of motion is obtained from the action in the same manner as was used for the KG equation of motion. The Dirac action is given by

$$
(SD)_{g(x)} = \left[e^{-\alpha(x)} \int e^{\alpha(y)}(\bar{\psi}i\gamma^\mu D_{\mu,y}\psi - m\psi)dy \right]_{g(x)}.
\tag{4.15}
$$

This action is obtained by parallel transport of the Dirac Lagrangian density at points, y, to a reference point, x. The action integral is defined because, after transport, the integrands are all scalar quantities in a single number structure, $\bar{C}_x^{g(x)}$.

Following the same argument as was used for the KG equation and minimizing with respect to $\bar{\psi}$ gives

$$
\delta S = \int e^{\alpha(y)}(\delta\bar{\psi}i\gamma^\mu D_{\mu,y}\psi - m\delta\bar{\psi}(y)\psi(y))dy = 0.
\tag{4.16}
$$

Since $e^{\alpha(y)}$ and $\delta\psi(\bar{y})$ are common factors for the integrand, one obtains

$$
i\gamma^\mu D_{\mu,y}\psi - m\psi = i\gamma^\mu(\partial_{\mu,y} + A_\mu(y))\psi - m\psi(y) = 0
\tag{4.17}
$$

as the Dirac equation of motion. The effect of the α field appears in the presence of the interaction between the $\vec{A}$ field and ψ. If a coupling constant is included into $D_{\mu,y}$ as in $D_{\mu,y} = \partial_{\mu,y} + cA_\mu(y)$, then the equation of motion becomes

$$i\gamma^\mu(\partial_{\mu,y} + cA_\mu(y))\psi - m\psi = 0. \tag{4.18}$$

4.4   Abelian gauge theory

Abelian gauge theories play an important role in physics. This is seen by the fact that quantum electrodynamics is a good example of an Abelian theory. These theories differ from the pure scalar case in the presence of an additional $U(1)$ gauge transformation that acts on the vector spaces $\bar{V}_y^{g(y)}$.

In the arena of local mathematics and no information at a distance, a vector field, β on the space–time, M is lifted to be a section on the fiber bundle, $\mathfrak{M}\mathfrak{F}^g$. For each y in M, $\beta(y)$ becomes the vector $\psi(y)_{g(y)}$ in $\bar{V}_y^{g(y)}$. The vector value of the field at y is $\psi(y)$.

The Dirac Lagrangian of Eq. (4.14) will be used as an illustrative example of Abelian gauge theory. The connection appearing in the definition of the derivative,

$$D_{\mu,y}\psi = \frac{C_g(y, y + d_\mu y)\psi(y + d_\mu y)_{g(y+d_\mu y)} - \psi(y)_{g(y)}}{d_\mu y}, \tag{4.19}$$

is expanded to include a $U(1)$ gauge factor as in

$$\begin{aligned} C_g(y, y + d_\mu y) &= V(y, y + d_\mu y)U(y, y + d_\mu y) \\ &= (1 + A_\mu(y)d_\mu y)(1 + iB_\mu(y)d_\mu y) \\ &= 1 + (A_\mu(y) + iB_\mu(y))d_\mu y. \end{aligned} \tag{4.20}$$

This equation is valid to first order in small factors. Both factors in the last line of the equation express the fact that

$$V(y, y + d_\mu y) = e^{\alpha(y+d_\mu y) - \alpha(y)} = 1 + A_\mu(y)d^\mu y$$

and $U(y, y + d_\mu y) = 1 + iB_\mu(y)d^\mu y$.

The presence of $V(y, y + d_\mu y)$ in the connection follows from the observation that the scaling factor for the vector space at each location in M depends on the location. Instead of the usual action of the unitary gauge operator given by

$$U(y, y + d_\mu y)\bar{V}_{y+d_\mu y} = \bar{V}_y,$$

one has

$$U(y, y + d_\mu y)\bar{V}_{y+d_\mu y}^{g(y+d_\mu y)} = \bar{V}_y^{g(y+d_\mu y)}. \tag{4.21}$$

This vector space does not belong to any fiber in the fiber bundle for the used setup of local mathematics and no information at a distance. This is fixed by the component $V(y, y+d_\mu y)$ of the connection. One has

$$V(y, y + d_\mu y)\bar{V}_y^{g(y+d_\mu y)} = Z_y^{y+d_\mu y}\bar{V}_y^{g(y+d_\mu y)} = \frac{g(y + d_\mu y)}{g(y)}\bar{V}_y^{g(y)}. \tag{4.22}$$

The $Z_y^{y+d_\mu y}$ map is given by Eqs. (1.113) and (1.114).

The action of $V(y, y + d_\mu y)$ on vectors can be obtained from Eq. (1.107). Replacement of the subscripts s and t with y and $y+d_\mu y$ and s and t with $e^{\alpha(y)}$ and $e^{\alpha(y+d_\mu y)}$ gives $\psi_t = \psi(y + d_\mu y)_{y+d_\mu y}$, $\psi_s = \psi(y)_y$, $s_s = [e^{\alpha(y)}]_y$, and $t_s = [e^{\alpha(y+d_\mu y)}]_y$. Use of these results in Eq. (1.107) gives

$$\psi(y + d_\mu y)_{y+d_\mu y} \equiv [e^{\alpha(y+d_\mu y)-\alpha(y)}]_y \psi(y + d_\mu y)_y$$
$$= V(y, y + d_\mu y)\psi(y + d_\mu y)_{y+d_\mu y}. \tag{4.23}$$

The fact that the connection can be represented as a component of the gauge group $GL(1, R+) \times U(1)$ is another expression of the solution to the problem outlined above. This representation requires that both $V(y, y + d^\mu y)$ and $U(y, y + d^\mu y)$ be present as factors of the connection (Figure 4.1). Here, $GL(1, R+)$ is the one-dimensional positive real general linear group.

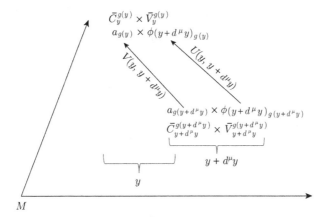

Figure 4.1. Action of scalar and vector transport factors in the connection from $y + d^\mu y$ to y. The figure clearly shows that if the scalar factor $V(y, y + d^\mu y)$ was deleted from the connection, the constant a would be in the fiber at $y + d^\mu y$ and the vector ϕ in the fiber at y. The distance between neighboring points, y and $y + d^\mu y$, in space–time, M, is greatly magnified for illustration purposes.

Use of Eq. (4.20) for the connection in Eq. (4.19) gives

$$D_{\mu,y}\psi = (\partial_{\mu,y} + A_\mu(y) + iB_\mu(y))\psi. \qquad (4.24)$$

The Dirac Lagrangian density becomes

$$\mathcal{L}\mathcal{D}(y) = \bar{\psi}i\gamma^\mu\big(\partial_{\mu,y} + cA_\mu(y) + ieB_\mu(y)\big)\psi - m\bar{\psi}\psi. \qquad (4.25)$$

The subscript $g(y)$ has been suppressed. Coupling constants, c and e, for the $\vec{A}$ and $\vec{B}$ fields have been added. If $\vec{B}$ is the photon field, then e is the electron charge.

A basic requirement of Abelian theories is that the Lagrangians are invariant under transformations by both global and local $U(1)$ transformations. Invariance under a global transformation where $\psi(y) \to e^{i\phi}\psi$ is obvious since $e^{i\phi}$ is a constant. This is not the case for local transformations where

$$\psi(y) \to U(y)\psi(y) = e^{i\phi(y)}\psi(y). \qquad (4.26)$$

This is seen by the observation that

$$\partial_{\mu,y}(U(y)\psi(y)) = (\partial_{\mu,y}U(y))\psi(y) + U(y)\partial_{\mu,y}\psi(y).$$

Invariance can be expressed by the condition [16,45]

$$D'U(y)\psi = U(y)D\psi, \qquad (4.27)$$

where

$$U(y) = e^{i\phi(y)}. \qquad (4.28)$$

Use of the fact that

$$\partial_{\mu,y}U(y) = U(y)i\partial_{\mu,y}\phi \qquad (4.29)$$

gives the result that

$$U(y)(\partial_{\mu,y} + cA'_\mu(y) + eiB'_\mu(y) + i\partial_{\mu,y}\phi)\psi$$
$$= U(y)(\partial_{\mu,y} + cA_\mu(y) + ieB_\mu(y))\psi. \qquad (4.30)$$

Separating out the real and imaginary parts gives

$$A'_\mu(y) = A_\mu(y),$$
$$B'_\mu(y) = B_\mu(y) - \partial_{\mu,y}\phi/e. \qquad (4.31)$$

One concludes from this that invariance of terms of the Lagrangian under local $U(1)$ gauge transformations has no effect on the mass of the $\vec{A}$ field. It can have any mass including 0. The same invariance condition means that the mass of the $\vec{B}$ field must be 0.

In addition to the mass restriction differences, the $\vec{A}$ and $\vec{B}$ fields differ in another way. The field, $\vec{A}$, is integrable as it is the gradient of a scalar field α. As a result, for points y and x separated by a finite distance, $V(x,y)$, as an integral along a path p parameterized by s with $y = p(0)$ and $x = p(1)$, is given by

$$V(x,y) = e^{-\int_0^1 |\vec{A}(p(s))|ds} = e^{-\int_0^1 A_\mu(p(s))\frac{dp^\mu(s)}{ds}ds} = e^{\alpha(y)-\alpha(x)}. \qquad (4.32)$$

This result follows from the equality

$$A_\mu(p(s))\frac{dp^\mu(s)}{ds} = \frac{d\alpha(p(s))}{dp^\mu(s)}\frac{dp^\mu(s)}{ds} = \frac{d}{ds}\alpha(p(s)). \qquad (4.33)$$

One sees from this that $V(x,y)$ is independent of the path between y and x. The value of $V(x,y)$ is given by the exponential of the difference in values of the integral at the endpoints.

The parallel transport operator for the photon field, $\vec{B}$, is different in that $\vec{B}$ is nonintegrable. This operator is given by [16]

$$U_p(x, y) = e^{-\int_0^1 B_\mu(p(s)) \frac{dp^\mu(s)}{ds} ds}. \tag{4.34}$$

For this field, there is no equation equivalent to Eq. (4.33) for $\vec{B}$. Physically, the nonintegrability of this field is responsible for the Aharonov–Bohm effect [47].

The QED Lagrangian density with the $\vec{A}$ field and a mass term for $\vec{A}$ is obtained by adding a dynamical term for the photon field. It is given by [45]

$$\mathcal{LQ}(y) = \bar{\psi} i \gamma^\mu (\partial_{\mu,y} + c A_\mu(y) + ie B_\mu(x)) \psi$$
$$- m\bar{\psi}\psi - \frac{1}{4} B^{\mu,\nu} B_{\mu,\nu} - \frac{1}{2} m_A A^\mu(y) A_\mu(y). \tag{4.35}$$

The great accuracy of the QED Lagrangian, without the presence of the $\vec{A}$ interaction with the field, ψ, for electrically charged particles, means that the terms containing A_μ must be very small. This requires that either the coupling constant, c, must be very small compared to the fine structure constant or that $\vec{A}(y)$ must be very close to 0 in regions of space and time where the great accuracy of QED has been or can be verified.

As noted, α is a real scalar field with gradient field, $\vec{A}$. Any mass is possible. In addition, since there is no reason to assign a nonzero spin to the field, it can be assumed to be a spin 0 field.

There are several candidate fields in physics whose known properties are consistent with these requirements. These include dark energy, dark matter, the inflaton, the Higgs boson, and quintessence. Intriguing connections of the α field with these candidates include the fact that the Higgs field is also a spin 0 scalar field. If the coupling constant, c, is very small, this would be a property shared with dark matter as a WIMP. Also, as will be seen in a later chapter, the α field can account for the red shift resulting from the Hubble expansion and dark energy.

In other somewhat related work on the relativity of arithmetics [48,99], a possible connection to dark energy has been suggested. These connections will be discussed in more detail in the last part of this book.

The equation of motion for the QED Lagrangian is an extension of that for the Dirac Lagrangian. Equation (4.35), with the $g(y)$ subscript added, is parallel-transported to a reference point, x. The action is given by

$$SQ_{g(x)} = \int C_g(x,y)[\mathcal{L}\mathcal{Q}(y)dy]_{g(y)} = \left[e^{-\alpha(x)} \int e^{\alpha(y)} \mathcal{L}\mathcal{Q}(y))dy \right]_{g(x)}$$

$$= \left[e^{-\alpha(x)} \int \left(e^{\alpha(y)} \bar{\psi} i\gamma^\mu (\partial_{\mu,y} + cA_\mu(y) + ieB_\mu(x))\psi \right. \right.$$

$$\left. \left. - m\bar{\psi}\psi - \frac{1}{4} B^{\mu,\nu} B_{\mu,\nu} - \frac{1}{2} m_A A^\mu(y) A_\mu(y) \right) dy \right]_{g(x)} . \quad (4.36)$$

Minimization of the action to obtain the equation of motion is essentially the same as that for the Dirac equation, so the steps will not be given. The resulting equation of motion, with constants c and e included, is

$$i\gamma^\mu D_{\mu,y}\psi - m\psi = i\gamma^\mu (\partial_{\mu,y} + cA_\mu(y) + ieB_\mu(y))\psi - m\psi = 0. \quad (4.37)$$

4.5   Non-Abelian gauge theory

The treatment of non-Abelian gauge theory is an extension of that for Abelian theory. For the simplest case of SU(2) theory, the fields ψ are pairs or doublets of Dirac fields [46] as in

$$\psi = \begin{pmatrix} \psi_1 \\ \psi_2 \end{pmatrix}. \quad (4.38)$$

This theory was first developed by Yang and Mills [25] for fields as isospin doublets for proton and neutron fields.

As was the case for the Abelian theory, the usual derivative, $\partial_{\mu,y}\psi$, appearing in the Lagrangian, is not defined. Since $\psi(y + d^\mu y)$ and $\psi(y)$ are in different vector spaces, $\bar{V}_{y+d^\mu y}^{g(y+d^\mu y)}$ and $\bar{V}_y^{g(y)}$, the implied subtraction is not defined. This is fixed by the use of a connection to parallel transform $\psi(y + d^\mu y)$ to a state in $\bar{V}_y^{g(y)}$.

As was the case for the Abelian gauge theory, the connection consists of two factors as in

$$C_g(y, y + d^\mu y) = V(y, y + d^\mu y)U(y, y + d^\mu y). \qquad (4.39)$$

The definition of the scalar part, $V(y, y + d^\mu y)$, remains the same as it is in the Abelian theory. The definition of $U_{y,y+d^\mu y}$ is different as it is an element of the group $SU(2)$.

For points $y + d^\mu y$ in the neighborhood of y, the expression of $U_{y,y+d^\mu y}$, in terms of the Pauli matrices as generators of the Lie algebra $su(2)$, is given by

$$U_{y,y+d^\mu y} = e^{-iE_\mu^j(y)\frac{\sigma_j}{2}d^\mu y}. \qquad (4.40)$$

The sum is over all three j values, the σ_j are the Pauli matrices, and the $\vec{E}^j$ for $j = 1, 2, 3$ are a triplet of vector fields.

The covariant derivative of ψ has the same form as that for the Dirac Lagrangian, as in Eq. (4.19). It is given by

$$D_{\mu,y}\psi = \frac{C_g(y, y + d_\mu y)\psi(y + d_\mu y)_{g(y+d_\mu y)} - \psi(y)_{g(y)}}{d_\mu y}. \qquad (4.41)$$

Equation (4.39) and expansion of $V(y, y + d^\mu y)$ and $U_{y,y+d^\mu y}$ to first order in small quantities give

$$D_{\mu,y}\psi = \left(\partial_{\mu,y} + aA_\mu(y) - igE_\mu^j(y)\frac{\sigma_j}{2}\right)\psi. \qquad (4.42)$$

Coupling constants, a and g, have been added for the terms showing $G(1, R+)$ and $SU(2)$ fields interacting with ψ.

The requirement that the terms in the Lagrangian be invariant under local $SU(2)$ transformations,

$$U(w) = e^{-iw^j(x)\cdot\frac{\sigma_j}{2}}, \qquad (4.43)$$

is given in Eq. (4.27). Here, the field ψ in

$$D'_\mu U\psi = UD_\mu\psi \qquad (4.44)$$

is a doublet of fields. Also, U is obtained from Eq. (4.43).

For the $\vec{A}$ field, Eq. (4.44) gives the result that

$$A'_\mu = A_\mu. \tag{4.45}$$

This is a consequence of the fact that $\vec{A}$ as the gradient of the scalar field, α, does not interact with the local gauge transformation. This is expressed by the commutation relation

$$[\vec{A}, U(w)] = 0. \tag{4.46}$$

The requirement of Eq. (4.44) on the three $\vec{E}^j$ fields is not affected by the presence of the $\vec{A}$ field. It is given by the well-known result [45,46],

$$E_\mu^{j'} = E_\mu^j + \varepsilon^{jkl} w^k E_\mu^l - \frac{1}{g}\partial_\mu w^j. \tag{4.47}$$

The gauge-invariant Lagrangian is obtained in a fashion similar to that for the Abelian case. A gauge-invariant dynamical term for the $\vec{E}$ fields is constructed from the second-rank antisymmetrical tensor [45]

$$F_{\mu,\nu}^j = \partial_\mu E_\nu^j - \partial_\nu E_\mu^j + g\varepsilon^{jkl} E_\mu^k E_\nu^l. \tag{4.48}$$

The resulting Lagrangian is given by

$$\mathcal{L} = \bar{\psi}i\gamma^\mu D_\mu \psi - m\bar{\psi}\psi - F_{\mu,\nu}^j F^{j,\mu,\nu} - \frac{1}{2}m_A A_\mu A^\mu. \tag{4.49}$$

Here, D_μ is given by Eq. (4.42) as

$$D_\mu = \partial_\mu + aA_\mu - igE_\mu^j\frac{\sigma_j}{2}. \tag{4.50}$$

Gauge invariance has the result that the masses of the $\vec{E}$ fields are all 0.[3] A mass term is present for the $\vec{A}$ field as gauge invariance imposes no restrictions on the mass.

---

[3]These fields, as vector bosons, acquire mass as a result of the Higgs mechanism [45,46].

4.6   Complex value fields

The emphasis of this book is on real, integrable vector fields, A, that are gradients of a scalar real-valued field α. The limitation to real-value fields seems especially appropriate for geometry. There, the scalar field associated with different geometries is the real number structure.[4] Also, restriction to an integrable value field greatly simplifies the description of connections and their use.

However, nothing prevents one from expanding the value field to be complex. This is based on the existence of complex value factors for complex numbers. As shown in Section 1.7 of Chapter 1, there are an infinite number of complex number structures, each with a different complex value factor. Each of the structures satisfies the axioms for complex numbers.

So far, the real vector field A has been restricted to be integrable as the gradient of the value field α. If the value field is complex, as in $\alpha + i\beta$, then the gradient vector field has two components as in $A + iB$. In this section, a brief outline of some of the effects of a complex value field and the removal of the integrable restriction on none, one, or both of the vector field components will be investigated.

Complex value fields are appropriate for use in gauge theories and quantum mechanics. This is based on the fact that the vector spaces appearing in these theories include complex scalars in their axiomatic description. Also, in quantum electrodynamics, the photon field is nonintegrable. It therefore seems worthwhile to briefly outline properties of complex vector fields in which none, one, or both the real and imaginary components are nonintegrable.

If either the real or complex parts, or both, of the scaling field are nonintegrable, the resulting connections between points depend on the path between the points [16]. A connection $C_g(x, y)$ for integrable value fields must be replaced by a path-dependent connection $C_p(x, y)$. For a complex vector value field $A(z) + iB(z)$, the connection for a local mathematical structure $\bar{S}_y^{g(y)}$ with value factor $g(y)$ at y along a path from y to x is given by

$$C_p(x, y)\bar{S}_y^{g(y)} = \bar{S}_x^{g_p(x,y)}. \qquad (4.51)$$

---

[4]Complex geometries, i.e., those spaces with an associated complex scalar structure, have been discussed in [24].

Here, $g_p(x, y)$ is the numerical path-dependent scale change factor given by

$$g_p(x, y) = e^{-\int_0^1 (A_\mu(p(s)) + iB_\mu(p(s)))dp^\mu(s)}. \tag{4.52}$$

The path p is parameterized by s with $p(0) = y$ and $p(1) = x$.

Equation (4.52) is used to define the action of the connection, $C_p(x, y)$, as the multiplication of a number by the number with number value, $g_p(x, y)$. This is shown by parallel transport of numbers $a_{g(y)}$ in $\bar{S}_y^{g(y)}$ to numbers in $\bar{S}_x^{g_p(x,y)}$. One has

$$C_p(x, y)a_{g(y)} = [g_p(x, y)a]_x. \tag{4.53}$$

From this equation and Eq. (4.52), one sees that $C_p(x, y)$ can be split into the product of a real and imaginary component as in

$$C_p(x, y) = V_p(x, y)W_p(x, y). \tag{4.54}$$

The two factors are defined from Eq. (4.52) as

$$V_p(x, y) = e^{-\int_0^1 (A_\mu(p(s)))dp_s^\nu} \tag{4.55}$$

and

$$W_p(x, y) = e^{-\int_0^1 iB_\mu(p(s)))dp_s^\nu}. \tag{4.56}$$

If A is integrable, then the A integral in the exponent is replaced by $\alpha(y) - \alpha(x)$. If iB is integrable, the B integral is replaced by $i\beta(y) - i\beta(x)$. Here, iB is the gradient of the $i\beta$ value field.

The connections satisfy a path composition property. If p is a path and q is a path whose beginning point, x, coincides with the endpoint of p, then

$$C_{q*p}(z, y) = C_q(z, x)C_p(x, y). \tag{4.57}$$

The path $q * p$ is the concatenation of q with p and z is the endpoint of q.

One consequence of the path dependence of the connection is that the fiber bundle, $\mathfrak{M}\mathfrak{F}^g$, used for real integrable value fields is no longer appropriate. The reason is that the association of a fiber, F_y, with structures scaled by a single value factor, $g(y)$, is not tenable.

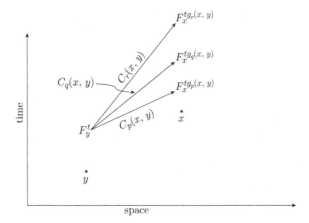

Figure 4.2. Illustration of the path dependence of the parallel transport in space–time of the fiber, F_y^t, to fibers at x. The dependence is shown for three paths, $p, q,$ and r. The point x is timelike relative to y.

The path dependence of the connections means that for a given set of structures in a fiber at y, each with a value factor, t, there are many structures in a fiber at x, each of which is the result of parallel transport of the structure at y to x. Each of the parallel transports is implemented by a connection along a specific path. The existence of many structures at x that are possible parallel transports of the structure with value factor t at y is a consequence of the path dependence of the connections.

Figure 4.2 illustrates the path dependence of parallel transport of structures with value factor t at y to structures at x for three different paths, p, q, and r. The structures in a fiber at y or x are denoted collectively with F replacing a specific structure.

A fiber bundle that can accommodate path-dependent parallel transport is a simple product bundle

$$\mathfrak{MF} = M \times F, \pi, M. \tag{4.58}$$

For gauge theories, the fiber F is the collection over all complex scaling factors, t, of complex number structures and vector spaces suitable for gauge theories, as in

$$F = \cup_t \bar{C}^t \times \bar{V}^t. \tag{4.59}$$

The projection operator π is defined by $\pi(x, F) = x$ with inverse $\pi^{-1}(x) = F_x$, where

$$F_x = \cup_t \bar{C}_x^t \times \bar{V}_x^t. \tag{4.60}$$

The fiber, F_x, is roomy enough to accommodate path-dependent parallel transports from any structure pair, $\bar{C}_y^t \times \bar{V}_y^t$, to a structure pair in the fiber at x. Also, if desired, the fiber can be expanded to include other mathematical structures that might be relevant to the problem under consideration.

Accommodation of the path-dependent connections into the derivatives for the Dirac Lagrangian appearing in Abelian gauge theories is straightforward. The derivatives for a matter field, given by Eq. (4.19), are

$$D_{\mu,y}\psi = \frac{C_g(y, y + d_\mu y)\psi(y + d_\mu y)_{g(y+d_\mu y)} - \psi(y)_{g(y)}}{d_\mu y}. \tag{4.61}$$

The derivative becomes

$$D_{\mu,y} = \partial_{\mu,y} + A_\mu(y) + iB_\mu(y). \tag{4.62}$$

The Dirac Lagrangian becomes

$$\begin{aligned}
\mathcal{LD}(y) &= \bar{\psi}(y)(i\gamma^\mu D_{\mu,y}\psi) - m\bar{\psi}(y)\psi(y) \\
&= \bar{\psi}(y)i\gamma^\mu(\partial_{\mu,y} + cA_\mu(y) + eiB_\mu(y))\psi - m\bar{\psi}(y)\psi(y). \tag{4.63}
\end{aligned}$$

Coupling constants c and e have been added.

The requirement that the Lagrangian be invariant under all local $U(1)$ transformations as in

$$D'_{\mu,y}U(y)\psi = U(y)D_{\mu,y}\psi \tag{4.64}$$

gives the result

$$A'_\mu(y) = A_\mu(y) \tag{4.65}$$

and

$$B'_\mu(y) = B_\mu(y) - \partial_{\mu,y}\phi/e. \tag{4.66}$$

As was the case in Eq. (4.31), $U(y) = e^{i\phi(y)}$.

From these relations, one concludes that there are no restrictions on the mass of the A field. Any mass is possible. The mass of the B field must be 0.

4.6.1 *Is A the Higgs particle and $\vec{B}$ the photon?*

There are four integrable possibilities for the A and B fields in that none, either one, or both fields can be integrable. If the B field is nonintegrable, it may be possible to identify it with the photon. The observation that the mass of the imaginary iB field is zero and the covariant derivative has the form shown in Eq. (4.62) is consistent with this possibility.

In this book, the A field is assumed to be integrable. As such, it is the gradient of a real field α. These conditions are consistent with the possible identification of α with the Higgs field [52]. Also, both the Higgs field and α are both spin 0 real scalar fields. Additionally, like the Higgs, α can have mass.

These identifications of A with the Higgs and iB with the photon are highly provisional. Strong support or refutation for these identifications requires further work. Also, the use of a complex value field in other areas of physics and in geometry and as a value field for local mathematical structures would need more work. Some work in this direction on gauge theory in complex geometry has been published earlier [24].

If A and B can be identified with the Higgs particle and the photon, it would imply a deep and close connection between physics and local mathematics at a foundational level. Perhaps Tegmark [10] is right: "physics is mathematics". Also, "mathematics is physics".

The reason that the B field was not included in other chapters of Part II is that it did not seem relevant to the application of local mathematics and connections to geometric properties. For these properties, the relevant connections are real and not complex.

Chapter 5

Quantum Mechanics

5.1 Introduction

Quantum mechanics is another area of physics where the effects of local mathematics and the value field affect theoretical descriptions of properties of quantum systems [51]. Several examples of these properties are described in the following sections of this chapter. Quantities to be discussed include wave functions of one or more systems, including entangled states. Hamiltonian, momentum and energy, and quantum dynamics are also described.

Quantum mechanics adds something new that was not present in the description of gauge theories. This is the fact that wave packets have two equivalent complementary descriptions: one as a wave function over coordinate space and the other as a wave function over momentum space. If scaling is present in the description based on coordinate space, it is also present in the description based on momentum space. This can be seen by taking the Fourier transform of the space localized wave packet.

The plan of this chapter is to describe nonrelativistic quantum mechanics in three-dimensional Euclidean space. The quantum mechanics of one particle are described first. Included are discussions of wave packets, momenta, Hamiltonians, and expectation values. The description is then extended to include quantum dynamics with time evolution. This material is repeated for the quantum mechanics of two particles. The description is brief because emphasis is placed on the new aspects that are not present for one-particle systems. Entanglement is one example. There is a brief discussion of the quantum mechanics of n particles. The description for n particles is brief because nothing essentially new is added.

The last section consists of an extension of local mathematics and the presence of a value field to momentum space. A few properties of quantum systems in this representation are discussed.

5.2 One particle

Quantum mechanical descriptions of states and properties of one particle are a good place to show the effects of the g field. Here nonrelativistic quantum mechanics is treated with particle states and properties described in three-dimensional Euclidean space. A very brief outline of the usual description is presented first. Mathematical systems that make no reference to space or time are used for the outline. These are global systems in that they are "outside" of space and time. The effect of the g field arises in the transition of these descriptions to those based on local mathematical systems. These are mathematical systems associated with different points in space.

5.2.1 *Usual description*

Let $\psi(y)$ be the wave function for a single spinless particle. For each y, $\psi(y)$ is the amplitude for finding the particle in state ψ at y. For each y, $\psi(y)$ is a complex number in a single complex number structure, $\bar{C}$. The corresponding vector function, $\lambda(y) = \psi(y)|y\rangle$, is a vector in the Hilbert space, $\bar{H}$. Scalar products and expectation values for the particle, as $\langle \psi, \psi \rangle$ and $\langle \psi, \tilde{O}\psi \rangle$, are all numbers in $\bar{C}$ or $\bar{R}$. Here $\tilde{O}$ is any observable and $\bar{R}$ is the real number structure. If ϕ is another wave function, then the scalar product, $\langle \phi, \psi \rangle$, is a complex number in $\bar{C}$.

These and other expectation values and scalar products are all well defined because, for all y, $\psi(y)$ and $\phi(y)$ are numbers in a single complex number structure, and the $\lambda(y)$ are all vectors in a single Hilbert space, $\bar{H}$. For this reason, the spatial integral expressions of scalar products and expectation values are well defined.

5.2.2 *Description based on local mathematics*

The mathematical arena used here is different in that it has separate complex number and Hilbert space structures at each point of M as a three-dimensional Euclidean space. In this case, the wave functions

ψ and ϕ have values in the different local complex number structures. Similarly, the vector values of λ are vectors in different Hilbert spaces. Specifically, for each y, $\psi(y)$ and $\phi(y)$ are values of complex numbers in the local complex number structure $\bar{C}_y$ at y and $\lambda(y)$ is the value of a vector in $\bar{H}_y$. In the presence of the value field, g, the local complex number and Hilbert space structures are scaled structures. For each y, $\bar{C}_y$ becomes $\bar{C}_y^{g(y)}$ and $\bar{H}_y$ becomes $\bar{H}_y^{g(y)}$.

5.2.2.1   *Fiber bundles*

Fiber bundles are a useful tool for the representation of these states and other properties of quantum systems. They provide a unified mathematical arena for the description of many different properties.

The bundle already described is useful here. It is given by Eq. (2.12). It is repeated here as

$$\mathfrak{M}\mathfrak{F}^g = M \times F, \pi_g, M, \tag{5.1}$$

where

$$F = \cup_t F^t. \tag{5.2}$$

The union is over all positive real number values, t. The parameter t denotes a real number valued scaling or value factor. The projection operator π_g is defined so that the inverse selects the fiber with the scaling factor given by the value of g at the location under consideration. That is,

$$\pi_g^{-1}(x) = F_x^{g(x)}. \tag{5.3}$$

For nonrelativistic quantum mechanics, the base space M is a three-dimensional Euclidean space. Unless otherwise stated, this will be implicitly assumed.

For quantum systems, each fiber F^t contains a complex number structure and a Hilbert space, as in

$$F^t = \bar{C}^t \times \bar{H}^t. \tag{5.4}$$

The fiber at each location y is given by

$$\pi_g^{-1}(y) = F_y^{g(y)} = \bar{C}_y^{g(y)} \times \bar{H}_y^{g(y)}. \tag{5.5}$$

The fiber bundle with this description of fibers is the mathematical arena for description of wave packets and many of their properties.

The effect of g on the states arises in their representations as sections on the bundle. The wave functions ψ and ϕ are represented by the sections $\psi(-)_{g(-)}$ and $\phi(-)_{g(-)}$ on the bundle. For each y, $\psi(y)_{g(y)}$ and $\phi(y)_{g(y)}$ are numbers in $\bar{C}_y^{g(y)}$ with values $\psi(y)$ and $\phi(y)$. $\lambda(-)_{g(-)}$ is a vector section on the bundle where for each y $\lambda(y)_{g(y)}$ is the vector in $\bar{H}_y^{g(y)}$ with vector value $\lambda(y)$. These representations of ψ, ϕ, and λ as sections of scalar and vector fields on M will be used throughout.

5.2.3 *Some effects of localization*

The localization of number and Hilbert space structures and the representation of states, ψ, ϕ, and λ with components in complex number structures and Hilbert spaces at different locations cause problems. The reason is that scalar products and expectation values, expressed as integrals over M, are not defined. The addition of scalars and vectors, implied in the definition of integrals, is defined only within a structure. It is not defined between complex number structures at different locations or between Hilbert space structures at different locations.

There are two ways to remedy these problems. One method is to use connections to parallel transport $\psi(y)$, $\lambda(y)$, and $\phi(y)$ in different complex number and vector space structures to local structures at a reference point. One can then describe scalar products, expectation values, and other properties of the states within the single local structure. The disadvantage with this approach is that the scaling factor ratios that appear in these descriptions are such that the change of these descriptions at x to another location z retains a memory of x. This is problematic because one would expect that transport of the wave functions to x followed by transport to z should give the same outcome as direct transport to z.

The other method avoids these problems. It begins with each scalar product and expectation value of the original untransported wave functions and vectors. For each location, y, these quantities are considered to take numerical values in the local number and vector structures at y. Parallel transport of these numbers or vectors to a reference location x is followed by integration to obtain local representations of the values of the scalar products or expectation values. This method has the advantage that subsequent transport of the results at x to another point z erases all memory of x.

The difference between these two methods is a consequence of the previously noted fact that the order of carrying out arithmetic operations and parallel transport matters. Arithmetic multiplication implied in the definition of scalar products or expectation values does not commute with parallel transport.

In the following sections, these methods will be illustrated in more detail. Expectation values for the position and the momentum will be described. The momentum expectation value is an example for which the matrix elements are not diagonal in the position representation.

5.2.3.1 *Localization of wave functions and vectors*

As noted above, for each y, $\psi(y)_{g(y)}$ and $\phi(y)_{g(y)}$ are numbers in $\bar{C}_y^{g(y)}$ with values $\psi(y)$ and $\phi(y)$. Also $\lambda(y)_{g(y)}$ is a vector in $\bar{H}_y^{g(y)}$ with value $\lambda(y)$. A representation of $\lambda(y)_{g(y)}$ in terms of its components is given by

$$\lambda(y)_{g(y)} = \psi(y)_{g(y)} \odot_{g(y)} |y\rangle_{g(y)} = (\psi(y)|y\rangle)_{g(y)}. \qquad (5.6)$$

In this expression, $|y\rangle_{g(y)}$ is a vector in $\bar{H}_y^{g(y)}$ with vector value $|y\rangle$. Also $\odot_{g(y)}$ denotes scalar vector multiplication between $\bar{C}_y^{g(y)}$ and $\bar{H}_y^{g(y)}$.

The Hilbert space, $\bar{H}_y^{g(y)}$, as a local structure at y, is limited to contain lifts of quantum states, vectors, and numbers at location y in M. If $\psi(y)|y\rangle$ and $\phi(y)|y\rangle$ are two different vectors at location y in M, then $\bar{H}_y^{g(y)}$ contains the lifts, $\psi(y)|y\rangle_{g(y)}$ and $\phi(y)|y\rangle_{g(y)}$. The Hilbert space also contains linear combinations, $[\psi(y)|y\rangle + \phi(y)|y\rangle]_{g(y)}$, of these vectors and products of these vectors with scalars, such as $[b\psi(y)\rangle]_{g(y)}$. Here b is the value of a number $b_{g(y)}$ in $\bar{C}_y^{g(y)}$. It also contains quantities expressed as spatial integrals, such as expectation values, provided the integrands are parallel transported to y and then integrated.

If the states ψ and ϕ are sections, $\psi(-)_{g(-)}$ and $\phi(-)_{g(-)}$, on the fiber bundle, $\mathfrak{MF}^g$, of Eq. (5.1), then ψ and ϕ, as integrals of the wave functions $\psi(y)$ and $\phi(y)$ as integrals over M, are not defined. Similarly, scalar products and expectation values of observables for ψ and or ϕ as integrals over M are not defined.

One method of defining these quantities consists of parallel transport of all the numbers $\psi(y)_{g(y)}$, $\phi(y)_{g(y)}$ and vectors $\lambda(y)_{g(y)}$ to a reference location, x. The transport of these and other quantities from y to x is implemented by a connection, $C_g(x,y)$. The action of these number preserving value changing maps is defined by

$$(v_{g(x)}(\psi(y)_{g(y)}))_{g(x)} = C_g(x,y)\psi(y)_{g(y)} = (e^{\alpha(y)-\alpha(x)}\psi(y))_{g(x)},$$
$$(v_{g(x)}(\phi(y)_{g(y)}))_{g(x)} = C_g(x,y)\phi(y)_{g(y)} = (e^{\alpha(y)-\alpha(x)}\phi(y))_{g(x)},$$
$$(v_{g(x)}(\lambda(y)_{g(y)}))_{g(x)} = C_g(x,y)\lambda(y)_{g(y)} = (e^{\alpha(y)-\alpha(x)}\lambda(y))_{g(x)}.$$
$$(5.7)$$

In these equations, $\psi(y)_{g(x)}$ and $\phi(y)_{g(x)}$, as numbers in $\bar{C}_x^{g(x)}$, are different from the numbers $\psi(y)_{g(y)}$ and $\phi(y)_{g(y)}$. This is the case even though they have the same value. Similarly, $\lambda(y)_{g(x)}$ is a different vector in $\bar{H}_x^{g(x)}$ than is $\lambda(y)_{g(y)}$, even though they have the same vector values. The number $(v_{g(x)}(\psi(y)_{g(y)}))_{g(x)}$ is a number in $\bar{C}_x^{g(x)}$ that has the same value as $\psi(y)_{g(y)}$ has in $\bar{C}_x^{g(x)}$. Similar statements hold for $\phi(y)_{g(y)}$ and $\lambda(y)_{g(y)}$. Also $v_{g(x)}$ is the valuation function for numbers or vectors in $\bar{C}_x^{g(x)}$ or $\bar{H}_x^{g(x)}$. More details are given in Section 1.3.

Equation (5.7) shows that localization of the state, ψ, as a section on the bundle, to a localized state with all values in $\bar{C}_x^{g(x)}$, changes the state to a state ψ_x, where

$$\psi_x = e^{-\alpha(x)}\int e^{\alpha(y)}\psi(y)dy. \tag{5.8}$$

Here ψ_x is the value of the number, $(\psi_x)_{g(x)}$, in $\bar{C}_x^{g(x)}$. A similar equation holds for the localization of $\lambda(-)_{g(-)}$ to x. The result is represented by λ_x where

$$\lambda_x = e^{-\alpha(x)}\int e^{\alpha(y)}\lambda(y)dy = e^{-\alpha(x)}\int e^{\alpha(y)}\psi(y)|y\rangle dy, \tag{5.9}$$

λ_x is the value of the vector, $(\lambda_x)_{g(x)}$ in $\bar{H}_x^{g(x)}$. The effect of localization and the presence of the α field result in the presence of the exponential factors in this equation.

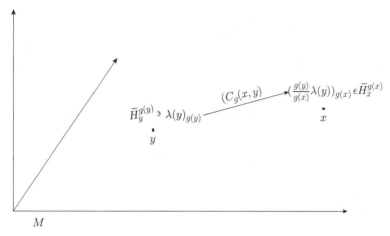

Figure 5.1. Illustration of parallel transport of the vector, $\lambda(y)_{g(y)}$ in $\bar{H}_y^{g(y)}$ at point y to the same vector in $\bar{H}_x^{g(x)}$. The transport is implemented by use of a connection, $C_g(x,y)$, as a vector preserving vector value changing map. In the fiber bundle representation for quantum mechanics, the connection maps all structures in the fiber at y to structures in the fiber at x. Illustrations for the parallel transport of the numbers, $\psi(y)_{g(y)}$ and $\phi(y)_{g(y)}$, are not given as they are the same as those for vectors. The only difference is that $\bar{C}_y^{g(y)}$ and $\bar{C}_x^{g(x)}$ replace $\bar{H}_y^{g(y)}$ and $\bar{H}_x^{g(x)}$.

Here the exponential representation of g as $g(y) = e^{\alpha(y)}$ is used. Either representation is valid. The g field form is more compact, so its use as a subscript is to be preferred.

Figure 5.1 illustrates the parallel transport of the vector $\lambda(y)_{g(y)}$ at y to the same vector at location x. Representations for ψ and ϕ are the same provided the Hilbert spaces are replaced by complex number structures.

5.2.3.2 *Scalar products of localized states*

Scalar products for the transported states localized in $\bar{H}_x^{g(x)}$ can now be defined. Use of Eq. (5.8) enables one to write

$$\langle \psi_x, \psi_x \rangle_{g(x)} = \langle (\psi_x)_{g(x)}, (\psi_x)_{g(x)} \rangle = \int \left[|(\psi_x(y)_{g(x)})|^2 \right]_{g(x)} dy_{g(x)}$$

$$= \int |C_g(x,y)(\psi(y)_{g(y)})|^2 dy_{g(x)}$$

$$= (e^{-2\alpha(x)})_{g(x)} \int (e^{2\alpha(y)})_{g(x)} |\psi(y)_{g(x)}|^2 dy_{g(x)}$$

$$= \left[e^{-2\alpha(x)} \int e^{2\alpha(y)} |\psi(y)|^2 dy \right]_{g(x)}. \tag{5.10}$$

In this equation, $\langle \psi_x, \psi_x \rangle_{g(x)}$ is a real number in $\bar{C}_x^{g(x)}$.[1]

This equation illustrates the use of square brackets with g values as subscripts to denote expressions as numbers or vectors with values enclosed by the square brackets. Parentheses are sometimes be used instead of square brackets.

Normalization of the state ψ_x requires that ψ_x be such that

$$\langle \psi_x, \psi_x \rangle_{g(x)} = 1_{g(x)}. \tag{5.11}$$

The question arises regarding the invariance of this normalization under change of reference point. To examine this, let z be another reference location. The normalization condition at z is given by

$$\langle \psi_z, \psi_z \rangle_{g(z)} = 1_{g(z)}. \tag{5.12}$$

The left-hand term of this equation can be obtained from Eq. (5.10) by substitution of z for x everywhere. Use of this in Eq. (5.12) gives

$$\left[e^{-2\alpha(z)} \int e^{2\alpha(y)} |\psi(y)|^2 dy \right]_{g(z)} = 1_{g(z)} \tag{5.13}$$

as an equivalent expression for the normalization condition at z.

Parallel transport of both sides of this equation from z to x gives

$$C_g(x, z) \left[e^{-2\alpha(z)} \int e^{2\alpha(y)} |\psi(y)|^2 dy \right]_{g(z)}$$

$$= \left[e^{-\alpha(z)-\alpha(x)} \int e^{2\alpha(y)} |\psi(y)|^2 dy \right]_{g(x)} \tag{5.14}$$

$$= C_g(x, z) 1_{g(z)} = [e^{\alpha(z)-\alpha(x)} 1]_{g(x)}.$$

Comparison of this with Eq. (5.11) shows that the normalization for $\psi_{g(z)}$ at z is different from the normalization condition for $\psi_{g(x)}$ at x.

---

[1]One can also expand the fibers to include local real number structures, $\bar{R}_y^{g(y)}$, at each point of M.

This shows that the normalization condition for a state is invariant under change of reference location from x to z if and only if $\alpha(z) = \alpha(x)$.

5.2.3.3 *Localization of scalar products of states*

In the above, scalar products of localized states were described. There is another way to obtain local descriptions of scalar products of states as sections over the fiber bundle. This begins with $[\phi^*(-)\psi(-)]_{g(-)}$ as a section on the bundle. Parallel transport of the values, $[\phi^*(y)\psi(y)]_{g(y)}$, to a reference location is followed by integration over y. This is described by an integral similar to that for Eq. (5.10) except that the connection acts outside of the absolute square.

For a reference location, x, one has

$$\langle\phi,\psi\rangle_{g(x)} = \int C_g(x,y)[\phi^*(y)\psi(y)]_{g(y)}\,dy_{g(y)}$$

$$= (e^{-\alpha(x)})_{g(x)} \int [e^{\alpha(y)}\phi^*(y)\psi(y)]_{g(x)}\,dy_{g(x)} \qquad (5.15)$$

$$= \left[e^{-\alpha(x)} \int e^{\alpha(y)}\phi^*(y)\psi(y)\,dy\right]_{g(x)}.$$

The value of this number in $\bar{C}_x^{g(x)}$ is enclosed in the square brackets. This result differs from that in Eq. (5.10) in that the scalar product is taken before parallel transport rather than after transport of the state. This difference is another example of the fact that taking products does not commute with the parallel transport operation.

Unlike the case for the scalar product of Eq. (5.10), parallel transport of $\langle\phi,\psi\rangle_{g(x)}$ to another reference point z gives the same result as direct localization of the scalar product to z. There is no memory of x in the transport to z. This follows from

$$C_g(z,x)\langle\phi,\psi\rangle_{g(x)}$$

$$= (e^{\alpha(x)-\alpha(z)})_{g(z)}(e^{-\alpha(x)})_{g(z)}\left[\int e^{\alpha(y)}\phi^*(y)\psi(y)\,dy\right]_{g(z)} \qquad (5.16)$$

$$= \langle\phi,\psi\rangle_{g(z)}.$$

Normalization of the state ψ, as in $\langle \psi, \psi \rangle_{g(x)}$, is preserved under change of reference location. This follows from

$$
\begin{aligned}
\langle \psi, \psi \rangle_{g(x)} = 1_{g(x)} &\to C_g(z,x) \langle \psi, \psi \rangle_{g(x)} \\
&= C_g(z,x) 1_{g(x)} \to \langle \psi, \psi \rangle_{g(z)} = 1_{g(z)}.
\end{aligned} \tag{5.17}
$$

As was seen above, norm preservation under reference point change from x to z is not the case for $\langle \psi_x, \psi_x \rangle_{g(x)}$. Instead one has $[\psi_x(z)]_{g(x)} = [e^{\alpha(z)-\alpha(x)}\psi(z)]_{g(x)}$.

Figure 5.2 illustrates the two methods, localization of matrix elements or matrix elements of localized states, of constructing local expressions for scalar products. It is shown for matrix elements of $\langle \psi, \psi \rangle$. It applies also to $\langle \phi, \psi \rangle$. Values of the section, $[\psi(-)]_{g(-)}$, on the fiber bundle are shown for two locations, y and z. The upper pair of horizontal arrows shows the results of first creating local matrix elements for the section at y and z, followed by parallel transport to x. The lower pair of horizontal arrows shows the results of parallel transport of the section values at y and z to x. These are used to construct local matrix elements for $\langle \psi_x | \psi_x \rangle$ at x.

5.2.3.4 *Localization of position expectation values*

The treatment of expectation values of observables that are diagonal in the position representation is similar to that for scalar products. An example is the localization of the expectation value for the position observable, $\tilde{Z}$. The usual expression for this is

$$
\langle \psi | \tilde{Z} | \psi \rangle = \int \psi^*(y) \psi(y) y \, dy. \tag{5.18}
$$

Replacement of ψ^* and ψ by their section representations, $\psi^*(-)_{g(-)}$ and $\psi(-)_{g(-)}$, on the fiber bundle gives

$$
\langle \psi | \tilde{Z} | \psi \rangle = \int \psi^*(y)_{g(y)} \psi(y)_{g(y)} y_{g(y)} \, dy_{g(y)} = \int [\psi^*(y) \psi(y) y \, dy]_{g(y)}. \tag{5.19}
$$

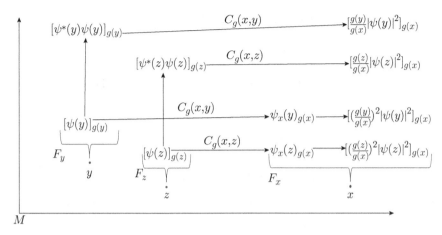

Figure 5.2. Illustration of the two methods of obtaining local integrand matrix elements for $\langle \psi | \psi \rangle$. Here $[\psi(y)]_{g(y)}$ and $[\psi(z)]_{g(z)}$ are the values of the fiber bundle section, $[\psi(-)]_{g(-)}$, at y and z. The vertical arrows show the construction of the local matrix elements for $\langle \psi | \psi \rangle$ at y and z. Parallel transport of these to the location x is shown by the horizontal arrows. They point to the matrix elements localized at x. The other method of creating matrix elements localized at x is shown by the horizontal arrows from $[\psi(y)]_{g(y)}$ and $[\psi(z)]_{g(z)}$. The arrows point to matrix elements of $\langle \psi | \psi \rangle$ generated at x from the parallel transport of $[\psi(y)]_{g(y)}$ and $[\psi(z)]_{g(z)}$ to x. The bundle fibers containing the different quantities shown in the figure at y, z, and x are indicated by the brackets labeled with F_y, F_z, and F_x.

The fact that this integral is not defined is fixed by parallel transport of the integrands to a common reference location, x. One obtains

$$
\begin{aligned}
\langle \psi | \tilde{Z} | \psi \rangle_{g(x)} &= \int C_g(x, y) [\psi^*(y)\psi(y) y \, dy]_{g(y)} \\
&= \left[e^{-\alpha(x)} \int e^{\alpha(y)} \psi^*(y)\psi(y) y \, dy \right]_{g(x)}.
\end{aligned}
\tag{5.20}
$$

5.2.3.5 *Position expectation values for localized states*

In the above localized expectation, values for the position observable have been determined. Alternatively, one can determine the expectation value for the position observable on localized states. For a reference point, x, use of Eq. (5.8) for the localized state gives

$$
\begin{aligned}
\langle \psi_x | \tilde{Z} | \psi_x \rangle_{g(x)} &= \left[e^{-2\alpha(x)} \int e^{2\alpha(y)} \langle \psi(y) \tilde{Z} \psi(y) \rangle dy \right]_{g(x)} \\
&= \left[e^{-2\alpha(x)} \int e^{2\alpha(y)} |\psi(y)|^2 y dy \right]_{g(x)} .
\end{aligned}
\tag{5.21}
$$

The difference between this and Eq. (5.10) for $\langle \psi_x, \psi_x \rangle$ is the replacement of dy with ydy.

Comparison of $\langle \psi_x | \tilde{Z} | \psi_x \rangle_{g(x)}$ in Eq. (5.21) with $\langle \psi \tilde{Z} | \psi \rangle_{g(x)}$ in Eq. (5.20) shows the presence of extra exponential factors in $\langle \psi_x | \tilde{Z} | \psi_x \rangle_{g(x)}$. This is another example of the fact that the product operation does not commute with parallel transport.

5.2.3.6   *Momentum expectation values for localized states*

Expectation values for observables that are not diagonal in the position representation are more complex. An example is the expectation value for the momentum observable, $\tilde{P}$. One has

$$
\langle \psi_x \tilde{P} \psi_x \rangle = \iint \psi_x^*(z) \langle z | \tilde{P} | y \rangle \psi_x(y) dz dy.
\tag{5.22}
$$

This equation is an equation in number values for numbers in $\bar{C}_x^{g(x)}$.

Equation (5.8) can be used to replace $\psi_x(z)$ and $\psi_x(y)$ by their expressions with scaling factors. One obtains

$$
\langle \psi_x | \tilde{P} | \psi_x \rangle = e^{-2\alpha(x)} \iint e^{\alpha(z)} \psi^*(z) \langle z | \tilde{P} | y \rangle e^{\alpha(y)} \psi(y) dz dy.
\tag{5.23}
$$

This equation shows that one difference between expectation values for diagonal and nondiagonal observables in the space representation is the replacement of $e^{2\alpha(y)}$ by $e^{\alpha(y)+\alpha(z)}$ in the space integral.

Another approach to the momentum expectation value is to first define the momentum representation of the state ψ_x. This is obtained as

$$\psi_x(p) = \frac{e^{-\alpha(x)}}{(2\pi\hbar)^{3/2}} \int e^{-ipy/\hbar} e^{\alpha(y)} \psi(y) dy. \qquad (5.24)$$

The momentum expectation value is then

$$\langle \psi_x | \tilde{P} | \psi_x \rangle = \int p |\psi_x(p)|^2 dp. \qquad (5.25)$$

Use of Eq. (5.24) to replace $\psi_x(p)$ in Eq. (5.25) gives

$$\langle \psi_x | \tilde{P} | \psi_x \rangle = \frac{e^{-2\alpha(x)}}{(2\pi\hbar)^3} \iiint p e^{ip(z-y)/\hbar} dp e^{\alpha(z)} \psi^*(z) e^{\alpha(y)} \psi(y) dz dy.$$

$$(5.26)$$

The presence of the exponential α factors distinguishes this expectation value from the usual one.

The momentum operator has a space representation as

$$\tilde{P}(y) = -i\hbar \frac{d}{dy} = -i\hbar \sum_{j=1}^{3} \partial_{j,y} = \sum_j P_j(y). \qquad (5.27)$$

Use of this representation gives

$$\tilde{P}\psi_x(y) = -i\hbar \frac{d}{dy} \psi_x(y) = -i\hbar \sum_{j=1}^{3} \partial_{j,y} \psi_x(y)$$

$$= e^{-\alpha(x)} \frac{-i\hbar}{(2\pi)^{3/2}} \sum_{j=1}^{3} \int \partial_{j,y} e^{ipy/\hbar + \alpha(y)} \psi(p) dp$$

$$= \frac{1}{(2\pi)^{3/2}} e^{-\alpha(x)+\alpha(y)} \sum_{j=1}^{3} \int (-i\hbar A_j(y) + P_j(y)) e^{ipy/\hbar} \psi(p)) dp$$

$$(5.28)$$

for the action of $\tilde{P}$ on the localized state, ψ_x at y. The component of the gradient of $\alpha(y)$ in the direction j is denoted by $A_j(y)$.

Equation (5.28) can be written in a simpler form as

$$\tilde{P}\psi_x(y) = -i\hbar \sum_j \partial_{j,y} e^{\alpha(y)} \psi(y) = -i\hbar e^{\alpha(y)} \sum_j (A_j(y) + \partial_{j,y})\psi(y).$$
(5.29)

This equation can be written in another form by first defining a complex momentum with components

$$P_{A,j} = P_j - i\hbar A_j.$$
(5.30)

Summation of the components in Eq. (5.30) gives

$$\vec{P}_A = \vec{P} - i\hbar\vec{A} = -i\hbar\nabla - i\hbar\vec{A}.$$
(5.31)

The momentum, $\vec{P}_A$, is the sum of the canonical momentum, $-i\hbar\nabla$, and the momentum, $-i\hbar\vec{A}$, associated with the alpha field.

The momentum expectation value for ψ_x is obtained from Eqs. (5.29) and (5.30) as

$$\langle\psi_x|\tilde{P}|\psi_x\rangle = e^{-2\alpha(x)} \iint e^{\alpha(z)} e^{\alpha(y)} \psi^*(z)\langle z|\vec{P}_A|y\rangle\psi(y)dzdy$$

$$= e^{-2\alpha(x)} \left\{ \iint e^{\alpha(z)} e^{\alpha(y)} \psi^*(z)\langle z|\vec{P}|y\rangle\psi(y)dzdy \right.$$

$$\left. -i\hbar \int e^{2\alpha(y)} |\psi(y)|^2 \vec{A}(y)dy \right\}.$$
(5.32)

5.2.3.7   *Localization of momentum expectation values*

So far local momentum expectation values have been determined by first using connections to localize states and then computing expectation values for the localized states. These methods are problematic. They are an example of the problems already shown in Section 2.4.1 for integrals of a function that is a product of two variables.

Momentum matrix elements, $\langle\psi(y)|\tilde{P}|\psi(z)\rangle$, cannot be factored into component products. Treatment of these elements as sections on a fiber bundle and parallel transport to a reference location uses the method outlined in Section 2.4.1 for functions $\theta(y, z)$ that cannot be factored.

For each variable pair y, z the matrix element is lifted to become $[\langle\psi(y)|\tilde{P}|\psi(z)\rangle]_{h(y,z)}$. This is a number in $\bar{C}_{y,z}^{h(y,z)}$. The pair value number $h(y, z)$ is defined by

$$h(y, z) = (g(y)g(z))^{1/2} = e^{(\alpha(y)+\alpha(z))/2}. \tag{5.33}$$

Parallel transport of $[\langle\psi(y)|\tilde{P}|\psi(z)\rangle]_{h(y,z)}$ to a reference location, x, is implemented by the connection $C_h(x, x; y, z)$. One obtains

$$
\begin{aligned}
&C_h(x, x; y, z)[\langle\psi(y)|\tilde{P}|\psi(z)\rangle]_{h(y,z)} \\
&= \left[\frac{h(y, z)}{h(x, x)}\langle\psi(y)|\tilde{P}|\psi(z)\rangle\right]_{h(x,x)} \\
&= \left[\frac{(g(y)g(z))^{1/2}}{g(x)}[\langle\psi(y)|\tilde{P}|\psi(z)\rangle]\right]_{g(x)}.
\end{aligned} \tag{5.34}
$$

The right-hand side of this equation is a number in $\bar{C}_x^{h(x,x)} = \bar{C}_x^{g(x)}$.

The integration of the integrand in Eq. (5.34) to obtain $[\langle\psi|\tilde{P}|\psi\rangle]_{g(x)}$ is defined because the integrands are all numbers in the same number structure. Also the result has the good property that parallel transport of $[\langle\psi|\tilde{P}|\psi\rangle]_{g(x)}$ to another location, w, gives $[\langle\psi|\tilde{P}|\psi\rangle]_{g(w)}$. This is the same result as would be obtained by using w instead of x as the first reference location.

The difference between $[\langle\psi|\tilde{P}|\psi\rangle]_{g(x)}$ obtained as an integral of the integrand of Eqs. (5.34) and (5.32) is another example of the lack of commutativity between parallel transport and arithmetic operations. In general, parallel transport of momentum expectation values is to be preferred because of the symmetry under change of reference locations.

The description given here and in the previous section is for the momentum operator. It also applies to other nondiagonal operators in general. The density operator with matrix elements, $\langle z|\tilde{\rho}|y\rangle$, is an example.

5.2.3.8 *The Schrödinger equation*

The effect of the α field on the momentum shows up in the kinetic energy term of the one particle Hamiltonian. The Hamiltonian is defined by

$$\tilde{H} = \tilde{K} + \tilde{V}. \tag{5.35}$$

$\tilde{K}$ and $\tilde{V}$ are the kinetic and potential energies.

With local mathematics as a basis, these operators act on the vector sections, $\psi(y)_{g(y)}$, in each Hilbert space, $\bar{H}_y^{g(y)}$. The operators are denoted by $\tilde{H}_{g(y)}$, $\tilde{K}_{g(y)}$, and $\tilde{V}_{g(y)}$.

For each y in M, the local Schrödinger equation is given by

$$\tilde{H}_{g(y)}\psi(y)_{g(y)} = (\tilde{K}_{g(y)} + \tilde{V}_{g(y)})\psi(y)_{g(y)} = E_{g(y)}\psi(y)_{g(y)}. \tag{5.36}$$

$E_{g(y)}$ is a real number in $\bar{C}_y^{g(y)}$ with value, E. The action of the kinetic energy operator on the state, $\psi(y)_{g(y)}$, is given by

$$\tilde{K}_{g(y)}\psi(y)_{g(y)} = \left(\frac{1}{2m}\right)_{g(y)} (\tilde{P}^2)_{g(y)}\psi(y)_{g(y)}$$

$$= \left(\frac{1}{2m}\right)_{g(y)} \sum_{j=1}^{3}(-i\hbar\partial_{j,y})^2\psi(y)_{g(y)}. \tag{5.37}$$

Here one has the same problem as was faced in defining derivatives for gauge theories described in Chapter 4. The derivative

$$\partial_{j,y}[\psi(y)]_{g(y)} = \lim_{dy_j \to 0} \frac{\psi(y+dy_j)_{g(y+dy_j)} - \psi(y)_{g(y)}}{(dy_j)_{g(y)}} \tag{5.38}$$

is not defined. The implied subtraction is between state components in different complex number structures, $\bar{C}_{y+d_jy}^{g(y+d_jy)}$ and $\bar{C}_y^{g(y)}$.

This problem can be fixed by following the method used for gauge theories. One parallel transports $\psi(y+dy_j)_{g(y+dy_j)}$ to the same number, $[(1+dy_j A_j(y))\psi(y)]_{g(y)}$, in the derivative. As before $A_j(y)$ is the j component of the gradient of α. The relation between $\partial_{j,y}$ and the value, $D_{j,y}$, of the derivative in the fiber at y is given by

$$\partial_{j,y}[\psi(y)]_{g(y)} = [D_{j,y}\psi(y)]_{g(y)} = [(A_j(y) + \partial_{j,y})\psi(y)]_{g(y)}. \tag{5.39}$$

The variable, y, outside the square brackets denotes a location in M. Inside it denotes the value or image of the location. As a result the derivative inside the brackets is defined.

The relation between the one particle kinetic energy operator acting on $[\psi(y)]_{g(y)}$ and the operator value, $\tilde{K}_y$, acting on the number value, $\psi(y)$, can be obtained from Eq. (5.39). This is given by

$$\tilde{K}_{g(y)}[\psi(y)]_{g(y)} = \left[-\frac{\hbar^2}{2m}\nabla_y^2\right]_{g(y)} [\psi(y)]_{g(y)} = \left[-\frac{\hbar^2}{2m}D_y^2\psi(y)\right]_{g(y)}$$

$$= \left[-\frac{\hbar^2}{2m}(\nabla_y + \vec{A}(y))^2\psi(y)\right]_{g(y)} = [\tilde{K}_y\psi(y)]_{g(y)}.$$

$$(5.40)$$

Expansion of the kinetic energy term in Eq. (5.40) gives

$$\tilde{K}_y\psi(y) = -\frac{\hbar^2}{2m}\sum_{j=1}^{3}(A_j(y) + \partial_{y,j})^2\psi(y)$$

$$= -\frac{\hbar^2}{2m}\sum_{j=1}^{3}[A_j(y)^2 + 2A_j(y)\partial_{j,y} + (\partial_{j,y}A_j(y)) + \partial_{j,y}^2]\psi(y).$$

$$(5.41)$$

Equation (5.41) shows that the kinetic energy of the scaled wave function is the sum of energies of the unscaled wave function, the energy, $-\hbar^2/2m\sum_{j=1}^{3}A_j^2(y)$, of the $\vec{A}$ field, and the interaction energy, $-\hbar^2/2m(\vec{A}(y)\cdot\nabla_y+\nabla_y\cdot\vec{A}(y))\psi(y)$, of the $\vec{A}$ field with the gradient function, ∇. There is also a contribution to the kinetic energy from the second derivative of α, as in

$$A_{j,j}(y) = \frac{\partial}{\partial y^j}A_j(y) = \left(\frac{\partial}{\partial y^j}\right)^2\alpha(y). \qquad (5.42)$$

The factor of two arises from the commutation relation:

$$\frac{\partial}{\partial y^j}(A_j(y)\psi(y)) = A_j(y)\frac{\partial}{\partial y^j}\psi(y) + \left(\frac{\partial}{\partial y^j}A_j(y)\right)\psi(y). \qquad (5.43)$$

The relation between the Schrödinger equation with the Hamiltonian as an operator acting on the number $\psi(y)_{g(y)}$ and as an operator value acting on the number value, $\psi(y)$, at y is given by

$$\tilde{H}_{g(y)}[\psi(y)]_{g(y)} = (\tilde{K}_{g(y)} + \tilde{V}(y)_{g(y)})[\psi(y)]_{g(y)}$$

$$= E_{g(y)}[\psi(y)]_{g(y)} \leftrightarrow [\tilde{H}_y\psi(y)]_{g(y)}$$

$$= [(\tilde{K}_y + \tilde{V}(y))\psi(y)_{g(y)}$$

$$= [E\psi(y)]_{g(y)} \leftrightarrow [\tilde{H}_y\psi(y) = (\tilde{K}_y + \tilde{V}(y))\psi(y)$$

$$= E\psi(y)]_{g(y)}. \tag{5.44}$$

The terms in Eq. (5.44) are all values of scalars or vectors in $\bar{C}_y^{g(y)}$ or $\bar{H}_y^{g(y)}$. The potential value, $\tilde{V}(y)$, is the value of the number $\tilde{V}(y)_{g(y)}$. The eigenvalue E is the number value of the number $E_{g(y)}$.

The limitation of the variable value y to be the image of the location in M appears to restrict the Schrödinger equation of Eq. (5.44) to one value, y. Parallel transport of the equation to a reference location, x, gives

$$C_g(x,y)[\tilde{H}_y\psi(y) = (\tilde{K}_y + \tilde{V}(y))\psi(y) = E\psi(y)]_{g(y)}$$

$$= [e^{\alpha(y)-\alpha(x)}]_{g(x)}[\tilde{H}_y\psi(y) = (\tilde{K}_y + \tilde{V}(y))\psi(y)$$

$$= E\psi(y)]_{g(x)} \tag{5.45}$$

or

$$[\tilde{H}_y\psi(y) = (\tilde{K}_y + \tilde{V}(y))\psi(y) = E\psi(y)]_{g(x)}. \tag{5.46}$$

Since the reference location is arbitrary, one can replace x by y and change the image variable to z in Eq. (5.44).

The presence of the vector field, $\vec{A}$, in the kinetic energy term makes Eq. (5.46) difficult to solve for eigenfunctions and eigenvalues, even for a simple potential, such as the harmonic oscillator.

There is another approach that throws some light on the solutions of this Schrödinger equation. So far the derivatives appearing in the Schrödinger equation have been defined for $\psi(y)_{g(y)}$. These derivatives, D_y, have been used to construct the equation which is then parallel transported to a reference location.

One can also parallel transport the wave function sections to a reference location, x, and define the Hamiltonian or Schrödinger equation for the transported wave function. The components of the wave function, transported to a reference location, x, are given by $[\psi_x(y)]_{g(x)}$, where

$$[\psi_x(y)]_{g(x)} = [e^{\alpha(y)-\alpha(x)}\psi(y)]_{g(x)}. \tag{5.47}$$

The Schrödinger equation for this state is given by

$$\left[\frac{-\hbar^2}{2m} \sum_j \frac{\partial^2}{\partial^2_{j,y}}\psi_x(y) + \tilde{V}(y)\psi_x(y) = E\psi_x(y) \right]_{g(x)}. \tag{5.48}$$

The partial derivatives are defined because the locations y and $y+d^j y$ are independent of x.

The equivalence between Eqs. (5.48) and (5.46) is seen from the following equivalences. One has

$$\frac{-\hbar^2}{2m} \sum_j \frac{\partial^2}{\partial^2_{j,y}}\psi_x(y) + \tilde{V}(y)\psi_x(y)$$

$$= E\psi_x(y) \Leftrightarrow \frac{-\hbar^2}{2m} \sum_j \frac{\partial^2}{\partial^2_{j,y}}e^{\alpha(y)-\alpha(x)}\psi(y)$$

$$+ \tilde{V}(y)e^{\alpha(y)-\alpha(x)}\psi(y)$$

$$= Ee^{\alpha(y)-\alpha(x)}\psi(y) \Leftrightarrow e^{\alpha(y)-\alpha(x)}\left(\frac{-\hbar^2}{2m}D_y^2\psi(y) + \tilde{V}(y)\psi(y) \right)$$

$$= e^{\alpha(y)-\alpha(x)}E\psi(y). \tag{5.49}$$

Cancellation of the common exponential factors on both sides of the equation gives the Hamiltonian in Eq. (5.46). This result assumes the potential commutes with the exponential factor.

The only difference between Eq. (5.48) and the usual representation of the Schrödinger equation is the replacement of ψ by ψ_x. This shows that the eigenvalues for ψ_x have the same number value as those for usual case with no value field present. The equations also show that eigenfunctions for the Hamiltonian of Eq. (5.48) have the same functional form as do the eigenfunctions for the Hamiltonian in the absence of the value field.

Note that, if the value factor is independent of the location, the representation of wave functions, potentials, and Hamiltonians as bundle sections has no effect on the values or properties of wave functions. In this case, the results obtained from solution of the Schrödinger equation are identical to those obtained in the usual case where wave functions, potentials, and Hamiltonians are treated as elements of global structures in global mathematics.[2]

5.2.4 *Quantum dynamics*

The results given in the previous sections describe the effect of the valuation field, α, on the theoretical description of quantum systems and properties in space. The descriptions, as mathematical expressions, are localized in space. Time played no role.

Expansion of the descriptions to include dynamics and time evolution in quantum mechanics in the presence of the α field can be done in the fiber bundle framework. The bundle, $\mathfrak{MF}^g$, of Eq. (5.1) is expanded to include time as part of the base space. The bundle becomes

$$\mathfrak{MTF}^g = M \times T \times F, \pi_g, M \times T. \tag{5.50}$$

The definition of the fiber remains the same as for $\mathfrak{MF}^g$ where Eqs. (5.2) and (5.4) give

$$F = \cup_s(\bar{C}^s \times \bar{H}^s). \tag{5.51}$$

Here s is any nonzero real number. The definition of π_g is expanded to include time as in

$$\pi_g^{-1}(x,t) = F_{x,t}^{g(x,t)} = \bar{C}_{x,t}^{g(x,t)} \times \bar{H}_{x,t}^{g(x,t)}. \tag{5.52}$$

The description of the value field is expanded to include time dependence as well as space dependence.[3]

---

[2]This is no longer true if one admits nonlinear value functions, as in the case of quantum mechanics formulated by means on non-Newtonian calculus [61].

[3]This extension to include time may seem excessive as it implies local mathematical structures at each point of space and time. However, this can be regarded as the nonrelativistic description of fiber bundles over M, where M is relativistic space–time as Minkowski space.

Wave functions for a single particle or system are treated here as sections, $\psi(-,-)_{g(-,-)}$, on $\mathfrak{MIF}^g$. For each y, t $\psi(y,t)_{g(y,t)}$ is a complex number in $\bar{C}_{y,t}^{g(y,t)}$ with value $\psi(y,t)$. This number value is the amplitude for finding the particle in state $\psi(y,t)_{g(y,t)}$ at y, t.

Time evolution of the wave function, ψ, is governed by the Schrödinger equation:

$$i\hbar \frac{d}{dt}\psi(y,t) = \tilde{H}(y)\psi(y,t). \tag{5.53}$$

The Hamiltonian, $\tilde{H} = \tilde{K} + \tilde{V}$, is assumed to be time-independent.

For ψ as a section on the fiber bundle, $\mathfrak{MIF}^g$, both the time derivative in Eq. (5.53) and the space derivative in the kinetic energy part of the Hamiltonian are not defined. For the space derivatives in the kinetic energy at a point y, t, this is fixed by use of Eq. (5.40) to write

$$\tilde{K}(y,t)\psi(y,t) = \frac{-\hbar^2}{2m} \sum_j D_{j,y,t}^2 \psi(y,t)$$

$$= \frac{-\hbar^2}{2m} \sum_j \left(\frac{\partial}{\partial y_j} + A_j(y,t)\right)^2 \psi(y,t). \tag{5.54}$$

Here

$$\frac{d\alpha(y,t)}{dy} = \vec{A}(y,t) = \sum_j A_j(y,t). \tag{5.55}$$

The Hamiltonian with $D_{j,y,t}$ replacing $\partial_{j,y}$ will be denoted by $\tilde{H}_D$.

The time derivative is obtained by use of the same methods that were used for the space derivative. For numbers, one obtains

$$\left(\frac{d}{dt}\right)\psi(y,t)_{g(y,t)} \rightarrow \left(\frac{d}{dt} + B(y,t)\right)_{g(y,t)} \psi(y,t)_{g(y.t)}$$

$$= \left[\left(\frac{d}{dt} + B(y,t)\right)\psi(y,t)\right]_{g(y,t)}. \tag{5.56}$$

In this equation,

$$B(y,t) = \frac{d}{dt}\dot{\alpha}(y,t), \tag{5.57}$$

where $g(y,t) = e^{\alpha(y,t)}$.

Use of Eqs. (5.54) and (5.56) enables one to use defined derivatives to express the Schrödinger equation for the wave function as a section on the fiber bundle. For each location, y, the equation is

$$\left[i\hbar \left(\frac{d}{dt} + B(y,t) \right) \psi(y,t) \right]_{g(y,t)} = [\tilde{H}_D(y)\psi(y,t)]_{g(y,t)}$$

$$= \left[\left(\frac{-\hbar^2}{2m} \sum_j D_{j,y}^2 + \tilde{V}(y) \right) \psi(y,t) \right]_{g(y,t)} . \tag{5.58}$$

This is an equation for numbers, vectors, and operators.

As was the case for the time-independent Schrödinger equation in Eqs. (5.44) and (5.46), Eq. (5.58) makes sense only if the space and time location variables are the same for the factors in the equation and for the value field, g. The reason is that, as a section on the fiber bundle, $\psi(z,u)_{g(z,u)}$ is a complex number in $\bar{C}_{z,u}^{g(z,u)}$ that has value $\psi(z,u)$.

As was the case for the time-independent Schrödinger equation, this section value restriction is removed by parallel transport of Eq. (5.58) to a common space and time reference location, x, s. The transported equation is given by

$$\left[i\hbar \left(\frac{d}{dt} + B(y,t) \right) \psi(y,t) \right]_{g(x,s)} = [\tilde{H}_D(y)\psi(y,t)]_{g(x,s)}$$

$$= \left[\left(\frac{-\hbar^2}{2m} \sum_j D_{j,y}^2 + \tilde{V}(y) \right) \psi(y,t) \right]_{g(x,s)} . \tag{5.59}$$

Integrals over space and or time are defined because the integration variables are independent of the reference location. The number value, vector value, and operator value equations are obtained by the removal of the $g(x,s)$ subscript. The result is

$$i\hbar \left(\frac{d}{dt} + B(y,t) \right) \psi(y,t) = \tilde{H}_D(y)\psi(y,t)$$

$$= \left(\frac{-\hbar^2}{2m} \sum_j D_{j,y}^2 + \tilde{V}(y) \right) \psi(y,t). \tag{5.60}$$

Replacement of $D^2_{j,y}$ by its equivalent from Eq. (5.40) in Eq. (5.60) gives

$$
i\hbar \left(\frac{d}{dt} + B(y,t) \right) \psi(y,t)
$$

$$
= \left(\frac{-\hbar^2}{2m} \sum_j (\partial_{j,y} + A_j(y))^2 + \tilde{V}(y) \right) \psi(y,t). \tag{5.61}
$$

Here $\partial_{j,y} = d/d_j(y)$.

The Schrödinger equation for the local state, ψ_x, that is equivalent to Eq. (5.60) or Eq. (5.61) is given by

$$
i\hbar \frac{d}{dt} \psi_{x,s}(y,t) = \left(\frac{-\hbar^2}{2m} \sum_j \partial^2_{j,y} + V(y) \right) \psi_{x,s}(y,t) = \tilde{H} \psi_{x,s}(y,t).
$$

$$\tag{5.62}$$

The relation between $\psi_{x,s}$ and ψ is given by

$$
\psi_{x,s}(y,t) = e^{-\alpha(x,s)+\alpha(y,t)} \psi(y,t). \tag{5.63}
$$

Assume that $\psi_{x,s}$ is an eigenfunction of $\tilde{H}$ with eigenvalue E. The equation expressing this is

$$
i\hbar \frac{d}{dt} \psi_{x,s}(y,t) = E \psi_{x,s}(y,t). \tag{5.64}
$$

Using Eq. (5.63) in Eq. (5.64) and taking the derivative give

$$
e^{-\alpha(x,s)+\alpha(y,t)} i\hbar \left(\frac{d}{dt} + B(y,t) \right) \psi(y,t) = e^{-\alpha(x,s)+\alpha(y,t)} E\psi(y,t).
$$

Removal of the exponents from both sides of the equation gives

$$
i\hbar \left(\frac{d}{dt} + B(y,t) \right) \psi(y,t) = E\psi(y,t). \tag{5.65}
$$

Rearrangement gives

$$
\frac{d}{dt} \psi(y,t) = -i \left(\frac{E - i\hbar B(y,t)}{\hbar} \right) \psi(y,t). \tag{5.66}
$$

This equation shows that if $\psi_{x,s}$ is an eigenfunction of $\tilde{H}$ with energy eigenvalue E, then the state ψ has a complex, time-dependent energy value of $E - i\hbar B(y,t)$.

This equation can be expressed in an integral form as

$$\psi(y,t) = e^{\int_{t_0}^{t}(iE/\hbar - B(y,s))ds}\psi(y,t_0). \qquad (5.67)$$

Use of the gradient theorem and the fact that $B(y,t)$ is the time derivative of α give the result that

$$\psi(y,t) = e^{-iE(t-t_0)/\hbar + \alpha(y,t_0) - \alpha(y,t)}\psi(y,t_0)$$

$$= e^{-i[E(t-t_0) + i\hbar(\alpha(y,t) - \alpha(y,t_0))]/\hbar}\psi(y,t_0). \qquad (5.68)$$

This result can also be obtained by noting that the time evolution of $\psi_{x,s}$ is given by

$$\psi_{x,s}(y,t) = e^{-iE(t-t_0)/\hbar}\psi_{x,s}(y,t_0). \qquad (5.69)$$

Replacement of $\psi_{x,s}$ by its equivalent expression in terms of ψ gives

$$e^{\alpha(y,t)-\alpha(x.s)}\psi(y,t) = e^{-iE(t-t_0)/\hbar}e^{\alpha(y,t_0)-\alpha(x,s)}\psi(y,t_0). \qquad (5.70)$$

Rearranging and canceling terms common to both sides of the equation give Eq. (5.68)

This result shows that the time dependence of the α field induces a time dependence of $\psi(y,t)$ that is in addition to the usual phase shift dependence. The presence of the real term, $\alpha(y,t)$, in the exponent means that the expectation value

$$\langle \psi(t), \psi(t) \rangle = \int \langle \psi(y,t), \psi(y,t) \rangle dy$$

$$= \int e^{2\alpha(y,t)-2\alpha(y,t_0)}|\psi(y,t)|^2 dy \qquad (5.71)$$

is time-dependent.

This shows that the norm of the state $\psi(t)$ is time-dependent. The time dependence is determined by fluctuations in the value of $\alpha(y,t)$ as t increases.

5.2.5   *Particle with spin*

The effect of number scaling on states of particles with spin is a simple extension of that for spinless particles. To see this, let ϕ be a wave packet state for a spin 1/2 particle. A general representation of such a state is given by $\phi(y)$, where

$$\phi(y) = \frac{1}{\sqrt{2}}(\psi_+(y)|+\rangle + \psi_-(y)|-\rangle) = \phi(y,+) + \phi(y,-). \quad (5.72)$$

This is an entangled state with entanglement between the space and spin degrees of freedom. It is a state in the tensor product space $\bar{H} \otimes \bar{S}$. Here $\bar{S}$ is a dimension-2 spin space with a spin-projection basis $|+\rangle$, $|-\rangle$. The expression for ϕ in Eq. (5.72) is a global state in the sense that the mathematical product space containing ϕ is not associated with any point of space, time, or space–time.

Here this setup is changed with local copies of the product space $\bar{H} \otimes \bar{S}$ at each point of M replacing the global product space. As before, M is three-dimensional Euclidean space. The fiber in the bundle, $\mathfrak{MF}^g$, is an extension of that for spinless particles. It is given by

$$F = \cup_t \bar{C}^t \times (\bar{H} \otimes \bar{S})^t. \quad (5.73)$$

The projection operator has a definition similar to that in $\mathfrak{MF}^g$. The fiber at each point y is

$$\pi_g^{-1}(y) = F_y^{g(y)} = \bar{C}_y^{g(y)} \times (\bar{H} \otimes \bar{S})_y^{g(y)}. \quad (5.74)$$

In this representation, the tensor products of the Hilbert space and spin space for a particle are localized at y with $g(y)$ the value factor. One can also represent the tensor product of the Hilbert space and spin space in Eq. (5.74) by

$$(\bar{H} \otimes \bar{S})_y^{g(y)} = \bar{H}_y^{g(y)} \otimes_{g(y)} \bar{S}_y^{g(y)}. \quad (5.75)$$

In this case, the tensor product operator is also scaled.

One proceeds in the same manner as was done for the spinless case. The state ϕ is represented as a section, $\phi(-)_{g(-)}$, on the fiber bundle. For each y, $\phi(y)$ is the value of the vector spin state, $\phi(y)_{g(y)} = [\frac{1}{\sqrt{2}}(\psi_+(y)|+\rangle + \psi_-(y)|-\rangle)]_{g(y)}$. This is a vector in $(\bar{H} \otimes \bar{S})_y^{g(y)}$.

As before, properties of $\phi(y)_{g(y)}$ that involve space integrals or derivatives over space are not defined. This can be fixed by parallel transport to a reference location x. This is represented by the action of a connection to give

$$\phi_x(y) = C_g(x,y)\phi(y)_{g(y)} = (e^{-\alpha(x)+\alpha(y)})_{g(x)}\phi(y)_{g(x)}$$
$$= [e^{-\alpha(x)+\alpha(y)}\phi(y)]_{g(x)} = \phi_x(y)_{g(x)}. \tag{5.76}$$

The wave packet integral,

$$\phi_{g(x)} = \int [e^{-\alpha(x)+\alpha(y)}\phi(y)]_{g(x)}dy_{g(x)}, \tag{5.77}$$

is defined as the integrands are all vectors in $(\bar{H} \otimes \bar{S})_x^{g(x)}$.

One sees from this that adding other degrees of freedom to the states of a particle, such as spin, has no effect on the valuation. This is the case even for space spin entangled states such as $\phi(y)$ in Eq. (5.72).

One can continue to investigate quantum properties of particles with spin in the arena of local mathematical structures with a value field present. This will not be done here as much of the treatment is similar to that for spinless particles.

5.3 Two particles

States of two or more particles can be described by the localization methods described so far. If the two particles are independent of one another, then the wave packet state of the two particles is just the product of the states of each particle. In this case, $\psi_{1,2} = \psi_1\psi_2$ and

$$\psi_{1,2}(y,z) = \psi_1(y)\psi_2(z), \tag{5.78}$$

for $j = 1,2$.

A fiber bundle framework that can be used to describe states as sections, $\psi_1(-)_{g(-)}\psi_2(-)_{g(-)}$, is the product of two fiber bundles as in

$$\mathfrak{MF}_{1,2}^g = \mathfrak{MF}_1^g \times \mathfrak{MF}_2^g. \tag{5.79}$$

On this fiber bundle, the two-particle state becomes $\psi_1(y)_{g(y)}\psi_2$ $(z)_{g(z)}$. Here $\psi_1(y)_{g(y)}$ and $\psi_2(z)_{g(z)}$ are numbers in $\bar{C}_y^{g(y)}$ and $\bar{C}_z^{g(z)}$.

Expectation values of this state for various observables are not defined. A particular example is the normalization integral:

$$\int |\psi_1(y)|^2_{g(y)} |\psi_2(z)|^2_{g(z)} dy_{g(y)} dz_{g(z)}$$

$$= \int |\psi_1(y)|^2_{g(y)} dy_{g(y)} \int |\psi_2(z)|^2_{g(z)} dz_{g(z)}. \qquad (5.80)$$

Parallel transport of the two states to a common reference location pair, x, x, fixes the problem. The normalization, $\langle \phi_1\phi_2|\phi_1\phi_2\rangle_{x,x}$, of the transported states

$$\phi_1(y)_{g(x)} = [e^{-\alpha(x)+\alpha(y)} \psi(y)]_{g(x)}$$

and

$$\phi_2(z)_{g(x)} = [e^{-\alpha(x)+\alpha(z)} \psi(z)]_{g(x)}$$

is defined. One obtains

$$\langle \phi_1\phi_2|\phi_1\phi_2\rangle_{g(x),g(x)}$$

$$= \left[e^{-4\alpha(x)} \int e^{2\alpha(y)+2\alpha(z)} |\psi_1(y)|^2 |\psi_2(z)|^2 dy dz \right]_{g(x),g(x)}. \qquad (5.81)$$

This description of the localized two-particle matrix element has the unsatisfactory property that parallel transport of the normalization to another location w, w preserves a memory of the localization to x, x. This problem is avoided by parallel transport of the matrix element, $|\psi_1(y)|^2_{g(y)} |\psi_2(z)|^2_{g(z)}$ to x, x. Parallel transport of the resulting matrix element,

$$\left[\frac{g(y)g(z)}{g(x)g(x)} |\psi_1(y)|^2 |\psi_2(z)|^2 \right]_{g(x),g(x)} \qquad (5.82)$$

$$= [e^{-2\alpha(x)+\alpha(y)+\alpha(z)} |\psi_1(y)|^2 |\psi_2(z)|^2]_{g(x),g(x)},$$

to w, w erases the memory of transport to x, x.

5.3.1 *Entangled states*

The above method of defining normalization and expectation values is limited to product states, $\psi_1\psi_2$. It does not work for two-particle entangled states that cannot be factored into the product of two single-particle states.

Examples of these states include Slater determinant states such as

$$\psi_{1,2}(y,z) = \frac{1}{\sqrt{2}}(\psi_1(y)\psi_2(z) - \psi_2(y)\psi_1(z)). \tag{5.83}$$

The subscripts $1, 2$ refer to the two particles and possibly to the values of the degree(s) of freedom entangled with space locations. An example of this type of state in which space and spin projection are entangled is the spin 0 state:

$$\psi_{+,-}(y,z) = \frac{1}{\sqrt{2}}(\psi_+(y)\psi_-(z) - \psi_-(y)\psi_+(z)). \tag{5.84}$$

Here $+$ and $-$ denote spin projections along and opposite some direction, respectively. This state gives the amplitude, $\psi_+(y)\psi_-(z) - \psi_+(z)\psi_-(y)$, for finding one spin projection at y and the other at z. The state has no information on which spin projection is where or on which particle is where.[4]

Inclusion of entangled states in the framework of local mathematics with the presence of the value field, α, can be done. The method for doing this is outlined in Section 2.4.1.

The fiber bundle, $\mathfrak{M}\mathfrak{Z}^h$, of Eq. (2.58) is useful here. The fiber, F, of the bundle is defined by

$$F = \cup_t \cup_S S^t. \tag{5.85}$$

The union over t includes all real scaling or value factors. The union over S can include mathematical systems of all types that include numbers in their description. Or it can be limited to the systems relevant to the problem under consideration. Here the systems in the fiber are limited to complex numbers and the product of two-particle Hilbert spaces, $\bar{H}_1 \otimes \bar{H}_2$.

---

[4]States which are entangled in spin only and not in space, such as $\psi_{1,2}(y,z) = \psi_1(y)\psi_2(z)B_{1,2}$, where $B_{1,2}$ is the Bell state, are treated as product states as far as the effects of the valuation field are concerned.

The inverse of the projection operator, π_h^{-1}, defines fibers containing the local mathematics at each location pair. For each pair, y, z, of locations in M,

$$\pi_h^{-1}(y, z) = F_{y,z}^{h(y,z)} = \bar{C}_{y,z}^{h(y,z)}, \ (\bar{H}_1 \otimes \bar{H}_2)_{y,z}^{h(y,z)}. \tag{5.86}$$

Note that the Hilbert space, $(\bar{H}_1 \otimes \bar{H}_2)_{y,z}^{h(y,z)}$, is not the same as $\bar{H}_y^{h(y,z)} \otimes \bar{H}_z^{h(y,z)}$. This is the tensor product of two Hilbert spaces, each at separate locations, whereas that of Eq. (5.86) is the tensor product of a pair of Hilbert spaces associated with the single location pair, y, z.

The definition of $h(y, z)$ is given in Eq. (2.65). It is repeated here as

$$h(y, z) = (g(y)g(z))^{1/2} = e^{(\alpha(y)+\alpha(z))/2}. \tag{5.87}$$

The wave function $\psi_{1,2}$ is represented by a section, $\psi_{1,2}(-, -)_{(h(-,-))}$, on the fiber bundle. For each location pair, y, z $[\psi_{1,2}(y, z)]_{h(y,z)}$ is a number in $\bar{C}_{y,z}^{h(y,z)}$.

Matrix elements of observables that are nondiagonal in the space representation, such as

$$[\psi_{1,2}^*(u, v)]_{h(u,v)} \tilde{O}[\psi_{1,2}(y, z)]_{h(y,z)},$$

are not defined if the arguments of h for $\psi_{1,2}^*$ differ from those of h for $\psi_{1,2}$. For diagonal observables, such as the identity, the normalization matrix element, $|\psi_{1,2}(y, z)|_{h(y,z)}^2$, is defined. However, the normalization integral,

$$\int |\psi_{1,2}(y, z)|_{h(y,z)}^2 dy dz_{h(y,z)},$$

is not defined.

This problem is fixed by use of a connection, $C_h(w, x; y, z)$, to parallel transport the matrix element to a common reference point pair, w, x. One has

$$|\psi_{1,2}(y, z)|^2_{h(y,z)} dydz_{h(y,z)}$$

$$\to C_g(w, x; y, z)|\psi_{1,2}(y, z)|^2_{h(y,z)} dydz_{h(y,z)}$$

$$= \left(\frac{h(y, z)}{h(x, w)} \right)_{h(x,w)} |\psi_{1,2}(y, z)|^2_{h(x,w)} dydz_{h(x,w)} \qquad (5.88)$$

$$= \left[\frac{h(y, z)}{h(x, w)} |\psi_{1,2}(y, z)|^2 dydz \right]_{h(x,w)}.$$

Within some general limits, to be discussed later, the choice of reference locations, x, w, is arbitrary. The point x can be quite close or quite far away from w. The order of the locations, x, then w, in x, w does not matter. The location pair x, w is the same as is w, x. The reference locations, x and w, can be different from one another or be the same. For the case where $x = w$ as x, x, the fiber

$$F^{h(x,x)}_{x,x} = F^{h(x,x)}_{x} = \bar{C}^{g(x)}_x \times (\bar{H} \otimes \bar{H})^{g(x)}_x. \qquad (5.89)$$

The location pair, x, x, is the same location in M.

Use of $h(x, x) = g(x)$ in the equation for the parallel transport of the local two-particle amplitude to x gives

$$\phi_{1,2}(y, z)_{h(x,w)} = C_h(x, w; y, z)\psi_{1,2}(y, z)_{h(y,z)}$$
$$= [e^{(-\alpha(x)-\alpha(w)+\alpha(y)+\alpha(z))/2}\psi_{1,2}(y, z)]_{g(x)}. \qquad (5.90)$$

Parallel transport of the normalization matrix element gives

$$|\phi_{1,2}(y, z)|^2_{g(x)} = [e^{-\alpha(x)+(\alpha(y)+\alpha(z))/2}|\psi_{1,2}(y, z)|^2_{g(x)}. \qquad (5.91)$$

Integration over y, z is defined because the matrix elements are all numbers in $\bar{C}^{g(x)}_x$.

5.3.2 *Momentum*

The results obtained so far can be used to describe the effect of the momentum operator and Hamiltonian on the two-particle state for the reference point pair, x, w. The action of the momentum operator on the local two-particle entangled state, $\phi_{1,2}(y,z)_{g(x,w)}$, in the fiber at x, w is given by

$$
\begin{aligned}
\tilde{P}_{h(x,w)}\phi_{1,2}(y,z)_{h(x,w)} &= [(\tilde{P}(y) + \tilde{P}(z))\phi_{1,2}(y,z)]_{h(x,w)} \\
&= [-i\hbar(\nabla_y + \nabla_z)\phi_{1,2}(y,z)]_{h(x,w)} \\
&= -[e^{-(\alpha(x)+\alpha(w))/2}i\hbar(\nabla_{g(y)} + \nabla_{g(z)}) \\
&\quad \times e^{(\alpha(y)+\alpha(z))/2}\psi_{1,2}(y,z)]_{h(x,w)} \\
&= -[e^{-(\alpha(x)+\alpha(w))/2}i\hbar(\nabla_y + \nabla_z) \\
&\quad \times e^{(\alpha(y)+\alpha(z))/2}\psi_{1,2}(y,z)]_{h(x,w)}.
\end{aligned}
\tag{5.92}
$$

The expression shows that the momentum of the two-particle state gives no information on the contribution of particles 1 and 2 to the momentum at y or at z. $\tilde{P}(y)$ is the momentum operator for the momentum at y, whichever particle is at y.

Commuting the gradients past the exponential of the α field gives

$$
\tilde{P}\phi(y,z)_{h(x,w)} = [e^{(-\alpha(x)-\alpha(w)+\alpha(y)+\alpha(z))/2}
$$

$$
\times(\tilde{P}_A(y) + \tilde{P}_A(z))\psi_{1,2}(y,z)]_{h(x,w)}.
\tag{5.93}
$$

The momenta, $\tilde{P}_A(y)$ and $\tilde{P}_A(z)$, include gradient fields of α as in

$$
\tilde{P}_A(y) = -i\hbar\sum_j\left(\partial_{j,y} + \frac{A_j(y)}{2}\right) = \vec{P}(y) - i\hbar\frac{\vec{A}(y)}{2},
\tag{5.94}
$$

$$
\tilde{P}_A(z) = -i\hbar\sum_j\left(\partial_{j,z} + \frac{A_j(z)}{2}\right) = \vec{P}(y) - i\hbar\frac{\vec{A}(y)}{2}.
\tag{5.95}
$$

Here $\tilde{P}(y)$ and $\tilde{P}(z)$ are one-particle momenta in the absence of the $\vec{A}$ field.

These results are similar to those obtained for the momentum of one particle. Here the effect of the $\vec{A}$ field is shared equally between the two particles.

To simplify notation, the subscript, x, w, common to both sides of Eq. (5.93) has been suppressed. It is understood that these equations are in terms of operator, vector, and number values of operators, vectors, and numbers in $(\bar{H}_1 \otimes \bar{H}_2)_{x,w}^{h(x,w)}$ and $\bar{C}_{x,w}^{h(x,w)}$ in the fiber at x, w.

5.3.3 *Energy*

These results can be used to determine the action of the two-body Hamiltonian on the two-particle state. The usual Hamiltonian for two particles with the same mass, m, is given by

$$\tilde{H}(y, z) = \frac{-\hbar^2}{2m} \sum_j \left(\frac{\partial^2}{\partial_{j,y}^2} + \frac{\partial^2}{\partial_{j,z}^2} \right) + V(y, z). \qquad (5.96)$$

With ψ as a section of the fiber bundle and with the value field, α, present, the partial derivatives in $\tilde{H}(y, z)$ must be replaced by $D_{j,y}$ and $D_{j,z}$. One has

$$\tilde{H}_D(y, z)_{h(y,z)} \psi(y, z)_{h(y,z)} = [\tilde{H}_D(y, z)\psi(y, z)]_{h(y,z)}$$

$$= \left[\left(\frac{-\hbar^2}{2m} \sum_j (D_{j,y}^2 + D_{j,z}^2) + V(y, z) \right) \psi(y, z) \right]_{h(y,z)}. \qquad (5.97)$$

Unless otherwise noted, the location subscript y, z will be dropped as it is understood that the Hamiltonian acts on the values of numbers and vectors in the fiber at y, z. This is shown in the second and third parts of the above equation. Note that the arguments of the derivatives and potential, y, z, are identical to those of the fiber location.

Expansion of the derivative, $D_{j,y}^2$, is given by

$$D_{j,y}^2 = D_{j,y} \left(\frac{\partial}{\partial_{j,y}} + \frac{A_j(y)}{2} \right)$$

$$= \left(\frac{\partial}{\partial_{j,y}} + \frac{A_j(y)}{2} \right) \left(\frac{\partial}{\partial_{j,y}} + \frac{A_j(y)}{2} \right)$$

$$= \frac{\partial^2}{\partial_{j,y}^2} + \frac{\partial}{\partial_{j,y}} \frac{A_j(y)}{2} + \frac{A_j(y)}{2} \frac{\partial}{\partial_{j,y}} + \frac{A_j(y)^2}{4} \qquad (5.98)$$

$$= \frac{\partial^2}{\partial_{j,y}^2} + \left(\frac{\partial}{\partial_{j,y}} \frac{A_j(y)}{2} \right) + A_{j(y)} \frac{\partial}{\partial_{j,y}} + \frac{A_j(y)^2}{4}.$$

One has a similar expression for $D_{j,z}^2$.

These results show that the two-particle kinetic energy is affected by the presence of the $\vec{A}$ field. The kinetic energy in the absence of $\vec{A}$ is given by $(-1/2m)(\tilde{P}^2(y) + \tilde{P}^2(z))$. In the presence of the $\vec{A}$ field, the kinetic energy is

$$
\begin{aligned}
KE(y,z)\psi(y,z) \\
= -\frac{1}{2m}(\vec{P}_A(y)^2 + \vec{P}_A(z)^2)\psi(y,z) \\
= -\frac{1}{2m}\left\{\left(\vec{P}(y) - i\hbar\frac{\vec{A}(y)}{2}\right)^2 + \left(\vec{P}(z) - i\hbar\frac{\vec{A}(z)}{2}\right)^2\right\}\psi(y,z) \\
= \frac{\hbar^2}{2m}\sum_j\left\{\left(\partial_{j,y} + \frac{A_j(y)}{2}\right)^2 + \left(\partial_{j,z} + \frac{A_j(z)}{2}\right)^2\right\}\psi(y,z).
\end{aligned}
$$

$$(5.99)$$

This result shows that the kinetic energy is real even though the contribution of the $\vec{A}$ field to the momentum is imaginary.

As Eq. (5.97) shows, $[\tilde{H}_D(y,z)\psi(y,z)_{h(y,z)}$ is valid only for the location y, z. Any combination of $[\tilde{H}_D(y,z)\psi(y,z)]_{h(y,z)}$ and $[\tilde{H}_D(u,v)\psi(u,v)]_{u,v}$ cannot be done as these quantities are in different fibers at different locations. As is the case for other quantities, this is fixed by parallel transport of $[\tilde{H}_D(y,z)\psi(y,z)]_{h(y,z)}$ to a common pair of reference locations, x, w. The result is given by

$$
[\tilde{H}_D(y,z)\psi(y,z)]_{h(y,z)} \to [e^{-\alpha(x,w)/2+\alpha(y,z)/2}\tilde{H}_D(y,z)\psi(y,z)]_{h(x,w)}.
$$

$$(5.100)$$

The relation between the Hamiltonian of Eq. (5.96) and that of Eq. (5.97) at x, w given by the commutation relation:

$$
[\tilde{H}(y,z)e^{\alpha(y,z)}]_{h(x,w)} = [e^{\alpha(y,z)}\tilde{H}_D(y,z)]_{h(x,w)}.
$$

$$(5.101)$$

The factor, $e^{-\alpha(x,w)}$, common to both sides of the equation, has been canceled.

The potential, $\tilde{V}(y,z)$ in $\tilde{H}_D$, in Eq. (5.100) is assumed to be a function that associates a numerical quantity, as a potential energy of interaction, for each pair, y, z of space locations. As such the functional form of $\tilde{V}(y, z)$ is not affected by the presence of the α field. This shown by

$$V(y, z)_{h(y,z)} \to C_g(x, w; y, z)V(y, z)_{h(y,z)}$$

$$= [e^{-\alpha(x,w)+\alpha(y,z)}V(y, z)]_{h(x,w)}. \tag{5.102}$$

This equation holds even if $\tilde{V}(y, z)$ is a function of the distance between y and z. The effect of α on distances between points, discussed in the following chapter, does not affect $\tilde{V}$. The reason is that, for all point pairs, y, z, in the base Euclidean space, the coordinate number triples for each point are real number triples in $\bar{R}^{g(x,w)}_{x,w}$.[5] $\tilde{V}(y - z)_{h(x,w)}$ is a numerical potential energy quantity in $\bar{R}^{g(x,w)}_{x,w}$ with value $\tilde{V}(y - z)$.

5.3.4 *Dynamics of two particles*

The dynamics or time evolution of two-particle quantum states is an extension of that for one-particle states. The fiber bundle is similar to that used for one-particle states. It is

$$\mathfrak{M}\mathfrak{T}\mathfrak{F}^g = M \times T \times F, \pi_g, M \times T. \tag{5.103}$$

Here T is the time. The fiber F is the same as that in Eq. (5.85) with the union over system types limited to the complex numbers and the product of Hilbert spaces for each of the particles.

As was the case for one-particle dynamics, the fields, h and α, are extended to depend on time as well as space. This is shown by

$$h(y, z, t) = \sqrt{g(y, t)g(z, t)} = e^{(\alpha(y,t)+\alpha(z,t))/2}. \tag{5.104}$$

The domain of the projection operator, π_h, is all of $M \times T$ and all of or a subset of F. It is best defined by its inverse which is

$$\pi_g^{-1}(y, z, t) = F^{h(y,z,t)}_{y,z,t} = \bar{C}^{h(y,z,t)}_{y,z,t} \times (\bar{H}_1 \otimes \bar{H}_2)^{h(y,z,t)}_{y,z,t}. \tag{5.105}$$

---

[5]This represents an expansion of the fibers in the bundle to include a real number structure.

In nonrelativistic quantum mechanics, time is a parameter, not a dynamical variable. For this reason, ψ is considered here to be a collection of sections on the fiber bundle with one section, $\psi_t = \psi(-,t)_{h(-,t)}$, for each time t. For each y, z, $\psi(y, z, t)_{y,z,t}$ is a vector in $(\bar{H}_1 \otimes \bar{H}_2)^{h(g(y,z,t))}_{y,z,t}$ with value $\psi(y, z, t)$.

The time-dependent Schrödinger equation for the two-particle state, $\psi(t)_{h(x,w,t)}$, is similar to that for the one-particle state. One has

$$[i\hbar]_{h(x,w,t)} \frac{d}{dt}(\psi(t)_{h(x,w,t)}) = [\tilde{H}_D \psi(t)]_{h(x,w,t)}. \qquad (5.106)$$

The fact that the time derivative,

$$\frac{d}{dt}(\psi(t)_{h(x,w,t)}) = \lim_{dt \to 0} \frac{\psi(t + dt)_{h(x,w,t+dt)} - \psi(t)_{h(x,w,t)}}{dt}, \qquad (5.107)$$

is not defined is remedied by the same method as was used for one-particle dynamics.

Extension of Eq. (5.56) to two particles gives

$$[i\hbar]_{h(x,w,t)}[D_{x,w,t}\psi(t)]_{h(x,w,t)} = [\tilde{H}_D \psi(t)]_{h(x,w,t)} \qquad (5.108)$$

as the time-dependent Schrödinger equation localized at the position pair, x, w, and time t. Here

$$[D_{x,w,t}\psi(t)]_{(x,w,t)} = \left[\left(\frac{d}{dt} + \frac{B(x,t) + B(w,t)}{2} \right) \psi(t) \right]_{h(x,w,t)}. \qquad (5.109)$$

The scalar potentials $B(x,t)$ and $B(w,t)$ are given by

$$B(x,t) = \frac{d}{dt}\alpha(x,t) \quad \text{and} \quad B(w,t) = \frac{d}{dt}\alpha(w,t). \qquad (5.110)$$

One would like to be able to integrate this equation to describe the state at different times based on the state at some reference time, s. This requires parallel transport of both sides of Eq. (5.108) to numbers and vectors in a reference fiber at x, w, s. The result of this transport multiplies both sides of the equation by the same exponential factor, $e^{-\alpha(x,w,s)+\alpha(x,w,t)}$. Cancellation of the exponential factors gives

$$i\hbar D_{x,w,t}\psi(t) = \tilde{H}_D\psi(t) \tag{5.111}$$

as an equation in number values, vector values, and operator values in structures at location, x, w, s. Here $\alpha(x, w, s) = (\alpha(x, s) + \alpha(w, s)/2$ and $\alpha(y, z, t) = (\alpha(y, t) + \alpha(z, t))/2$.

So far, the Schrödinger equation has been defined at time t and then parallel transported to a reference location in space pairs and time. One can proceed in the reverse direction by first transporting the state to a reference location and then forming the Schrödinger equation on the transported state. Two-particle quantum states that are parallel transported to a reference location, x, w, s, are expressed as in Eq. (5.90) with a time variable added. One has

$$\phi(t, y, z)_{h(x,w,s)} = [e^{-\alpha(x,w,s)+\alpha(x,w,t)}\psi(t, y, z)]_{h(x,w,t)}. \tag{5.112}$$

The subscripts $1, 2$ present in Eq. (5.90) have been suppressed.

Equation (5.111) for the state $\psi(t)_{h(x,w,s)}$ is equivalent to the usual time-dependent Shrödinger equation for the state $\phi(t)_{h(x,w,s)}$ localized at the position pair, x, w, and time, s. The equation is

$$i\hbar\frac{d}{dt}\phi(t)_{h(x,w,s)} = H\phi(t)_{h(x,w,s)}. \tag{5.113}$$

Suppose that $\phi(t)_{h(x,w,s)}$ is an eigenfunction of the Hamiltonian $\tilde{H}$ where

$$\tilde{H}\phi(t)_{h(x,w,s)} = E\phi(t)_{h(x,w,s)}. \tag{5.114}$$

The time-dependent Schrödinger equation is

$$i\hbar\frac{d}{dt}\phi(t)_{h(x,w,s)} = E\phi(t)_{h(x,w,s)}. \tag{5.115}$$

Use of the relation

$$\frac{\partial}{\partial t}\phi(t) = D_t\psi(t) \tag{5.116}$$

gives

$$i\hbar\left(\frac{d}{dt} + \frac{B(x,t)+B(w,t)}{2}\right)\psi(t) = E\psi(t)]. \tag{5.117}$$

This equation can be expressed as

$$\frac{1}{\psi(t)}\frac{d\psi(t)}{dt} = \frac{d\ln\psi(t)}{dt} = -\frac{B(x,t) + B(w,t)}{2} - \frac{i}{\hbar}E. \qquad (5.118)$$

The solution of this equation can be given in an integral form as

$$\psi(t) = \psi(t_0)\exp\left(\int_{t_0}^t \left(-\frac{B(x,u) + B(w,u)}{2} - \frac{i}{\hbar}E\right)du\right)$$

$$= \psi(t_0)\exp\left\{\int_{t_0}^t \left(-\frac{B(x,u) + B(w,u)}{2}\right)du - \frac{i}{\hbar}E(t - t_0)\right\}$$

$$= \psi(t_0)\exp\frac{-i}{\hbar}E(t - t_0)\exp\int_{t_0}^t \left(-\frac{B(x,u) + B(w,u)}{2}\right)du. \qquad (5.119)$$

The integrals are defined here because the integrands are all values of numbers and vectors in structures at the reference location, x, w, s.

As was shown for the single-particle case, one can use the gradient definition of $B(x,t)$ and $B(y,t)$ to simplify Eq. (5.119). The result is given by

$$\psi(t) = \psi(t_0)e^{-\frac{i}{\hbar}E(t-t_0)-\alpha(x,w,t)+\alpha(x,w,t_0)}. \qquad (5.120)$$

The same result is obtained from Eq. (5.115) by writing

$$\phi(t) = \phi(t_0)e^{-\frac{i}{\hbar}E(t-t_0)} \qquad (5.121)$$

and using Eq. (5.112) for the relation between $\phi(t)$ and $\psi(t)$.

5.4 *n*-particle states

The description of two-particle states extends easily to n-particle states. Fiber bundles and states, as sections on the bundles, are defined by replacement of pairs of space variables by n tuples of variables. A bundle is defined by

$$\mathfrak{M}\mathfrak{F}_{1,n}^h = M \times F, \pi_{h;1,\dots,n}, M. \qquad (5.122)$$

As before, M is the three-dimensional Euclidean space. The fiber is

$$F = \cup_c \bar{C}^c \times \bar{H}^c_{[1,n]}. \tag{5.123}$$

To save on notation $[1, n]$ denotes the particle numbers, $1, 2, \ldots, n$ The fiber at $y_{[1,n]} = y_1, y_2, \ldots, y_n$ is defined by an extension of π_h^{-1} in Eq. (5.86) from two locations to n locations:

$$\pi_h^{-1}(y_{[1,n]}) = F_{y_{[1,n]}}^{h(y_{[1,n]})} = \bar{C}_{y_{[1,n]}}^{h(y_{[1,n]})} \bigotimes \bar{H}_{y_{[1,n]}}^{h(y_{[1,n]})}. \tag{5.124}$$

The n locations $y_{[1,n]}$ are all in M. Here $\bar{H}_{[1,n]} = \bigotimes_{j=1}^{n} \bar{H}_j$ is the n-fold tensor product of the n single-particle Hilbert spaces.

One begins by representing an n-particle state as a section

$$(\psi_{[1,n]})_{h(-_{[1,n]})} = (\psi(-_{[1,n]})_{g(-_{[1,n]})} \tag{5.125}$$

to be a section on the fiber bundle, $\mathfrak{MF}_{1,n}^h$. For each n-point location, $y_{[1,n]}$, $\psi(y_{[1,n]})_{h(y_{[1,n]})}$ is an amplitude in $\bar{C}_{y_{[1,n]}}^{h(y_{[1,n]})}$ at $y_{[1,n]}$ with value $\psi(y_{[1,n]})|y_{[1,n]}\rangle$. As was the case for one- and two-particle states, the definition of integrals and derivatives of $(\psi_{y_{[1,n]}})_{h(y_{[1,n]})}$ requires prior transport to a reference n-point location, $x_{[1,n]} = x_1, x_2, \ldots, x_n$. The transport is implemented by an n-point connection $C_h(x_{[1,n]}, y_{[1,n]})$, where

$$(\psi_{y_{[1,n]}})_{h(y_{[1,n]})} \rightarrow C_h(x_{[1,n]}, y_{[1,n]})(\psi_{y_{[1,n]}})_{h(y_{[1,n]})}$$
$$= \frac{h(y_{[1,n]})}{h(x_{[1,n]})}(\psi_{y_{[1,n]}})_{h(x_{[1,n]})}. \tag{5.126}$$

The description of space and time integrals and derivatives for the transport of $(\psi_{y_{[1,n]}})_{h(y_{[1,n]})}$ to an n-point reference location, $x_{[1,n]}$, will not be given here as it is a simple extension of the description for two-particle states. The definition of $h(y_{[1,n]})$ is an extension to n points of that given for two particles. It is given by

$$h(y_{[1,n]}) = (g(y_1)g(y_2), \ldots, g(y_n))^{1/n} = e^{\alpha(y_{[1,n]})/n}. \tag{5.127}$$

Here

$$\alpha(y_{[1,n]}) = \sum_{j=1}^{n} \alpha(y_j). \tag{5.128}$$

One sees from this that $h(y_{[1,n]})$, as the geometric mean of the n single point g values, is equal to the exponential of the arithmetic mean of the n single point α values.

Chapter 6

Geometry

6.1 Introduction

In the previous chapters, the effects of the valuation field g on quantities in quantum mechanics and simple aspects of gauge theories were investigated. In these treatments, local Hilbert spaces and vector spaces, both with associated complex scalar structures, formed the basis of the treatment. The vector spaces were those appropriate for the description of different fields. Fiber bundles with M the base space provided a good framework for the discussion. The fiber contained a collection of scalar structures and vector or Hilbert spaces where each element in the collection corresponded to the scalars and vector spaces for a specific real value factor. The space M served as a background space.

For geometric properties and quantities, the overall framework based on local mathematics and no information at a distance remains the same as was used for quantum mechanics and gauge theories. It is useful to start with the product fiber bundle,

$$\mathfrak{M}\mathfrak{F}^g = M \times F, \pi_g, M, \tag{6.1}$$

that has been used so far. Here M is a flat space. Examples to be considered here consist of Euclidean space or space–time of special relativity. More general geometries will not be considered here.

For geometric properties, the contents of the fiber, F, need to be expanded to include representations of Euclidean space or space–time. This is done by writing

$$F = \cup_r (\bar{R}^r \cup \mathbb{S}^r \cup \ldots). \tag{6.2}$$

145

The union followed by dots indicates that the fiber may contain other structures besides representations of the real numbers and the spaces $\mathbb{S}^r$. Real numbers are emphasized because numerical geometric properties are mostly expressed in terms of real numbers.

The superscript r on $\mathbb{S}^r$ refers to the fact that numerical properties of structures in $\mathbb{S}^r$ and the geometry of $\mathbb{S}^r$ are based on the numbers in $\bar{R}^r$. The number, r, also serves as a scaling factor for the size of $\mathbb{S}^r$. For any small vector, $d\vec{y}_r$ in a Euclidean space, $\mathbb{S}^r$, the length is given by

$$|d\vec{y}|_r = [(r^2 \delta_{j,k} dy^j dy^k)^{1/2}]_r. \tag{6.3}$$

The size of the space with scaling factor r is encoded in the metric tensor, $[r^2 \delta_{j,k}]_r$. For space–time, the corresponding metric tensor is $[r^2 \eta_{\mu,\nu}]_r$. Here $\eta_{\mu,\nu} = \mathrm{diag}(1, -1, -1, -1)$.

For each r the space, $\mathbb{S}^r$ is a scaled image of M. Both are either Euclidean spaces or space–times. The relation between $\mathbb{S}^r$ and M is expressed by the use of charts [53]. For each value of r, a chart, ρ_r, is an open set preserving map of M to a representation, $\mathbb{S}^r$, of M, where

$$\mathbb{S}^r = \rho_r(M). \tag{6.4}$$

For Euclidean space, $\mathbb{S}^r$ is three-dimensional. It is four-dimensional for space–time. Since these spaces are defined to be images of M, the domain of ρ_r is all of M.

In the arena of local mathematics and the presence of the value field α, one defines the projection operator, π_g, in the fiber bundle by its inverse as in

$$\pi_g^{-1}(x) = F_x^{g(x)} = \bar{R}_x^{g(x)} \cup \mathbb{S}_x^{g(x)}. \tag{6.5}$$

Here $\mathbb{S}_x^{g(x)}$ is the local representation of M at x. The number structure, $\bar{R}_x^{g(x)}$, is used to describe all numerical properties of local geometric objects in the local space at x. As before

$$g(x) = e^{\alpha(x)}.$$

For Euclidean spaces, the replacement of r by $g(x)$ introduces a space dependence of the scaling or sizes of the local spaces. For space–times, the dependence may include time as well as space. The space and or time dependence of the scaling factor for the spaces is determined by the space and or time dependence of the field, $\alpha(x)$. The metric tensors for Euclidean space and space–time become $[e^{2\alpha(x)}\delta_{j,k}]_{g(x)}$ and $[e^{2\alpha(x)}\eta_{\mu,\nu}]_{g(x)}$.

The relationship between M and the local spaces extends to structures in M and their images in the local spaces, $\mathbb{S}^{g(x)}$. A path, p, or a membrane, m, in M has a corresponding image $p_{g(x)}$, or $m_{g(x)}$, in $\mathbb{S}_x^{g(x)}$.

The local images of structures in $\mathbb{S}_x^{g(x)}$ can be related to structures in M by the use of charts. This is done by appending either a location or value factor subscript on the chart map. For each x, the chart $\rho_{g(x)}$ maps structures in M to $\mathbb{S}_x^{g(x)}$. For a path p, one has

$$\rho_{g(x)}(p) = p_{g(x)}. \tag{6.6}$$

Replacement of p with m in this equation gives the same relation for membranes in M and their local images.

The effect of the local charts on paths in M is illustrated in Figure 6.1 for two sites, x and y, in M. The path and its local representations are shown by p, p_x, and p_y.

The relation between structures in M and their images in the local spaces extends to properties of structures in M and to images of their properties in the local spaces. For example, the length, $L(p)$, of a path in M has a corresponding image, $[L(p)]_{g(x)}$, in the local space at x.

Note that the local image of the path length in M is equal to the length of the image path in the space at x. That is,

$$[L(p)]_{g(x)} = [L(p_{g(x)})]_{g(x)}. \tag{6.7}$$

Here $L(p)$ is a number in the global real number structure and $[L(p_{g(x)})]_{g(x)}$ is a number in the local real number structure, $\bar{R}_x^{g(x)}$.

This lift of the whole path to a local space misses the main restrictions imposed by the "no information at a distance" principle and local mathematics. From gauge theory, one learns that vectors separated by infinitesimally small distances belong to separate vector spaces collocated with the vectors. This result is applied here to vectors in M. In particular, differential vectors in M are to be lifted to local spaces collocated with the vector. For example, the vector $d\vec{y}$ at

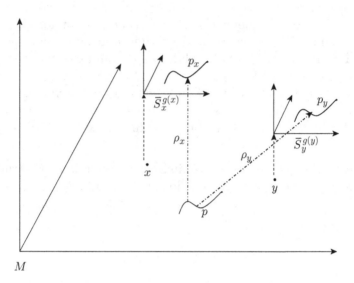

M

Figure 6.1. Local representations of a path p in M are shown. These are shown in the local spaces, $\bar{S}_x^{g(x)}$ and $\bar{S}_y^{g(y)}$, which are local representations of M. The local representations are implemented by charts, ρ_x and ρ_y. The space M and the local spaces are shown as three-dimensional Euclidean spaces. A similar representation holds for space–time.

y in M should be lifted to a corresponding differential vector $[d\vec{y}]_{g(y)}$ in $\mathbb{S}_y^{g(y)}$.

Implementation of this for extended objects such as paths in M requires that paths be represented by a bundle of tangent vectors. It is convenient to first represent a path, p, as parameterized by a real number s. For each s, the differential path tangent vector $d\vec{p}(s) = \vec{p}(s)ds$ in M is located at $p(s)$ in M. Lift of this vector to the collocated local space, $\mathbb{S}_{p(s)}^{g(p(s))}$, gives the local vector

$$[d\vec{p}(s)]_{g(p(s))} = [\vec{p}(s)ds]_{g(p(s))}$$

in $\mathbb{S}_{p(s)}^{g(p(s))}$.

Similar results apply to the lengths of vectors. If $L(d\vec{p}(s))$ is the length of $d\vec{p}(s)$ in M, the lifted length, as a numerical quantity in $\bar{R}_{p(s)}^{g(p(s))}$, is the number, $[L(d\vec{p}(s))]_{g(p(s))}$. The length value of this number is $L(p(s))$. The lifted length is the same number as is the length, $L([d\vec{p}(s)]_{g(p(s))})$, of the lifted vector. This equality is expressed by moving the length operation inside the square brackets as in

$$L([d\vec{p}(s)]_{g(p(s))}) = [L(d\vec{p}(s))]_{g(p(s))}. \qquad (6.8)$$

One now has the same problem as appeared in many other instances in previous chapters. Each of the small path tangent vectors is in a different local space. Assembling them into the tangent vectors of a path in one local space requires parallel transport of the vectors to a local space at a reference location for assembly. Similarly, assembling the differential lengths into a length integral requires parallel transport of the component lengths to a local space at some reference location. This is necessary because the integrands for the path length integral are numbers in different local real number structures. The integrand addition implied in the definition of the length integral is defined only within a single real number structure. It is not defined between numbers in different structures.

The rest of this chapter is concerned with the consequences of the use of parallel transport by connections to fix this problem. It turns out that for each point x in M, chosen as a reference space location, the collocated local space becomes distorted. It is no longer a flat space.

Local spaces at locations different from the reference location remain as flat spaces. They retain their scaling or location-dependent sizes.

From a mathematical point of view, the choice of reference locations is arbitrary. They can be anywhere in the cosmological universe. However, the locations of interest are those restricted to a very small fraction of the cosmological universe. The region includes the locations occupiable by us as observers.

A consequence of this is that charts mapping M to local space or space–times at reference locations should be replaced by an atlas [53] of charts. Each chart in the atlas maps an open subset of M to an open subset of $\mathbb{S}_x^{g(x)}$. The charts in the atlas differ in the scale factor they assign to the open set in their domain. This preserves the flatness of M. The other solution is to assume that M is not flat. In this case, $\mathbb{S}_x^{g(x)}$ is a faithful image of M.

The differential geometries on the local spaces described in this work are similar in some ways to that described by Weyl [49] almost 100 years ago. Weyl introduced a nonintegrable real vector field to describe the scaling under parallel transfer of quantities. The problem, as noted by Einstein [49], was that this required that the properties of measuring rods, clocks, and atomic spectra depend on their past history. He also noted that this contradicts the known physical properties of rods, clocks, and atomic spectra.[1]

There are significant differences between the geometries described here and those of Weyl. His geometry is described for M. Numerical values of quantities are based on a single real number structure. Local space–times with associated number structures and scaling factors depending on location in M play no role.

The construction described here is based on the use of local mathematics and the "no information at a distance principle" to construct local models of space or space–time. The geometries of the local reference models depend on space or space–time variations in the g or α field.

As was the case in Weyl geometry, values of properties of geometric objects in the local structures depend on the space or space–time location of corresponding structures (paths, etc.) in M. Einstein's criticisms are avoided by the restrictions on the space or space–time dependence of the value field. These are such that *local* deviations from flatness are too small to be observed experimentally. As will be seen later on, red shifts due to Hubble expansion and dark energy are examples of possible effects of value field variation that can show up at cosmological distances.

6.2   Vectors and vector values

Vectors are basic elements in M and in the scaled representations of M. It is therefore worthwhile to describe in some detail the relations between vectors in M and their images in the different image spaces. The description will be given for Euclidean space. It also applies equally to space–time. The relations will be given for the spaces in the fiber, F, in Eq. (6.2). This includes spaces for all scaling factors.

Let $\vec{v}$ be a vector in M. The image vectors in the local image spaces, $\mathbb{S}^r$ and $\mathbb{S}^s$, are denoted by $\vec{v}_r$ and $\vec{v}_s$. The vectors, $\vec{v}_r$ and $\vec{v}_s$, have the same vector value, $\vec{v}$, irrespective of whether $r = s$ or $r \neq s$.

---

[1]A good description of Weyl's work, including Einstein's remarks, and subsequent developments of early gauge theory are in the book by O'Raifeartaigh [50].

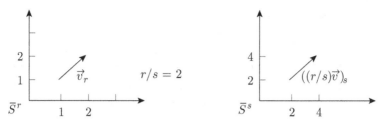

Figure 6.2. An illustration of the transport of a vector v_r in $\mathbb{S}^r$ to a vector in $\mathbb{S}^s$ for $r/s = 2$. The number preserving property is shown by the vector appearing the same in the two spaces. The fiducial marks in $\mathbb{S}^s$ show the value of the vector $\vec{v}_r$ is twice its value in the space $\mathbb{S}^r$.

Determination of whether two vectors, each in different spaces, are the same or not requires transport of the image vectors to a common space for comparison. This is implemented by a connection that maps $\vec{v}_r$ to $\mathbb{S}^s$ to become the vector $\vec{w}_s$ in $\mathbb{S}^s$. This vector is the same vector in $\mathbb{S}^s$ as $\vec{v}_r$ is in $\mathbb{S}^r$. It has a different vector value, $w = (r/s)v$, if and only if $r \neq s$.[2]

The distinctions between vector and vector value in different spaces are illustrated in Figure 6.2. As an aid to clarity, the spaces are shown as two-dimensional spaces, $\mathbb{S}^r$ and $\mathbb{S}^s$. The scaling ratio is given by $r/s = 2$. The relation between the vector in the space at the left and the one in the space at the right corresponds to a parallel transport of the vector. It shows that the length of the vector is preserved but the value of the length is increased by a factor of r/s. The increase in the length value is indicated by the smaller spacing of the fiducial marks in the space at the right compared to those in the space at the left.

The use of scaling or value factors to describe changes in the meaning or valuation of numbers in different real number structures and as point locations in local representations of M is similar in some ways to the concept of scaling used in physics and geometry. Scaling is not new, either in physics or in geometry. Scaling is an important component of renormalization theory [46], scale invariance of physical quantities, and of conformal field theories [54,55].

---

[2]Recall that parallel transports define the concept of "sameness" for elements in different spaces or number structures.

The number structure scaling or valuation used here is different from the types of scaling noted above. It differs from conformal transformations in that both angles between vectors and vector lengths are scaled. The scaling of angles may seem strange and counterintuitive. However, as will be seen, this does not cause problems. For example, the properties of number scaling or valuation are such that trigonometric relations are preserved under parallel transport from a fiber at one location to a fiber at another location.

The basic difference between the scaling used here and the other types of scaling used in physics is that both quantities and multiplication and division operations are scaled. The scaling must be such that the axiom validity of number and vector space structures is preserved under scaling. The type of scaling used here and in earlier work is a special case of that described in a recent paper in which functional relations between number structures of different types include the basic operations as well as the quantities [56].

6.3   Euclidean space

So far the relation between spaces and vectors in the spaces has applied to the fiber F in the fiber bundle. From now on, the relations will apply to spaces and structures in the fibers F_y at different locations, y, in M. The scale factors r and s are replaced by space-dependent factors $g(y) = \exp \alpha(y)$. The location dependence of the spaces is shown by their representation as $\mathbb{S}_y^{g(y)}$.

Most of the results to be obtained in this section apply to n-dimensional Euclidean space. However, three-dimensional space will be assumed unless otherwise mentioned. Properties of paths, as generalizations of vectors, are important structures in these and other types of spaces.

Let p be a path in M that is parameterized by a nonnegative real number s. The differential length elements of the path are given by

$$|d\vec{p}(s)| = (\nabla_s p \cdot \nabla_s p)^{1/2} ds. \tag{6.9}$$

The length of the path from $s = 0$ to $s = u$ is given by

$$L(p) = \int_0^u (\nabla_s p \cdot \nabla_s p)^{1/2} ds. \tag{6.10}$$

Two methods of creating local images of paths in local image spaces are considered. In one, the whole path is lifted to a local representation as an image path in a space at a reference location, In the other method, differential path tangent vectors are lifted to image vectors in image spaces at the same location as the tangent vectors. These are then parallel transported to a reference location for assembly into a path.

6.3.1 *Local images of whole paths*

A local image or representation in $\mathbb{S}_y^{g(y)}$ of path p in M can be obtained by use of a chart map $\rho_{g(y)}$ that maps structures in M into their images in $\mathbb{S}_y^{g(y)}$ in the fiber at y. The image path, $p_{g(y)}$, is defined by

$$p_{g(y)} = \rho_{g(y)}(p).\qquad(6.11)$$

The image of p at z is defined similarly by

$$p_{g(z)} = \rho_{g(z)}(p).\qquad(6.12)$$

Here $p_{g(y)}$ and $p_{g(z)}$ are paths in $\mathbb{S}_y^{g(y)}$ and $\mathbb{S}_z^{g(z)}$ that have path value p.

The points of the path $p_{g(y)}$ in $\mathbb{S}_y^{g(y)}$ are denoted by $p_{g(y)}(s)$. An equivalent representation of the image path point is by $p(s)_{g(y)}$. This is the path point in $\mathbb{S}_y^{g(y)}$ with path point value, $p(s)$.

The vector tangent to the path $p_{g(y)}$ at point s is denoted by $\nabla_s(p_{g(y)})$. The corresponding vector value is given by $\nabla_s p = \vec{p}(s)$.

The lengths of the paths as numbers in $\bar{R}_y^{g(y)}$ and $\bar{R}_z^{g(z)}$ are given by

$$\begin{aligned}
[L(p_{g(y)})]_{g(y)} &= \int_0^1 \sqrt{\nabla_s p_{g(y)} \cdot \nabla_s p_{g(y)}}\, ds_{g(y)} \\
&= \left[\int_0^1 \sqrt{\nabla_s p \cdot \nabla_s p}\, ds\right]_{g(y)} = L(p)_{g(y)}
\end{aligned}\qquad(6.13)$$

and

$$[L(p_{g(z)})]_{g(z)} = \int_0^1 \sqrt{\nabla_s p_{g(z)} \cdot \nabla_s p_{g(z)}} \, ds_{g(z)}$$

$$= \left[\int_0^1 \sqrt{\nabla_s p \cdot \nabla_s p} \, ds \right]_{g(z)} = L(p)_{g(z)}. \tag{6.14}$$

These equations have used the fact that the lift of the tangent vector for the path at $p(s)$ in M is the same as the tangent vector of the lifted path at $p(s)$ in the local image of M. That is,

$$\nabla_s p_{g(y)}(s) = \vec{p}_{g(y)}(s) = \rho_{g(y)}(\vec{p}(s)) = [\vec{p}(s)]_{g(y)} = [\nabla_s p(s)]_{g(y)}. \tag{6.15}$$

The path length values at the two locations, y and z, are given by the same value, $L(p)$. But $[L(p)]_{g(y)} \neq [L(p)]_{g(z)}$. Also the path, $p_{g(y)}$, is different from $p_{g(z)}$. These differences are a property of parallel transport of paths and path lengths from one reference location to another.

6.3.1.1 *Parallel transport of image paths*

Comparison of $p_{g(y)}$ with $p_{g(z)}$ is implemented by use of connections as in

$$q_{g(z)} = C_g(z, y) p_{g(y)}$$

$$= [e^{-\alpha(z)+\alpha(y)}]_{g(z)} p_{g(z)} = [e^{-\alpha(z)+\alpha(y)} p]_{g(z)}. \tag{6.16}$$

The path $q_{g(z)}$ is the same path in $\mathbb{S}^{g(z)}$ as $p_{g(y)}$ is in $\mathbb{S}^{g(y)}$. Their path values differ by the factor, $e^{-\alpha(z)+\alpha(y)}$.

One might object to the meaning of this equation as it includes the multiplication of a path by a number. This can be avoided by first approximating the path by a sum of short vectors tangent to the path at successive points. Figure 6.3 gives an illustration of this approximation for a path, $p_{g(y)}$, in $\mathbb{S}_y^{g(y)}$, in Euclidean space. The transported path can then be defined as the bundle of transported tangent vectors. This description is valid as vectors can be multiplied by numbers.

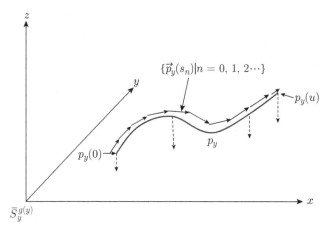

Figure 6.3. A representation of an approximation of a path $p_{g(y)}$ as a sequence of short tangent vectors. The subscript $g(y)$ is shortened to y. To avoid confusion, the vector sequence curve has been slightly lifted from the path. The vertical dashed arrows show the distance in the z direction of path points from the x, y plane.

The transport of the individual tangent vectors to location z results in a path with transported vectors defined by

$$\vec{q}(s)_{g(z)} = C_g(z, y)\vec{p}(s)_{g(y)} = [e^{-\alpha(z)+\alpha(y)}\vec{p}(s)]_{g(z)}. \tag{6.17}$$

The transported path, $q_{g(z)}$, is defined by the bundle of transported tangent vectors:

$$[\{\vec{q}(s) : 0 \le s \le u\}]_{g(z)} = [\{e^{-\alpha(z)+\alpha(y)}\vec{p}(s) : 0 \le s \le u\}]_{g(z)}. \tag{6.18}$$

The length of $q_{g(z)}(s)$ from $s = 0$ to $s = u$ is given by

$$L(q)_{g(z)} = \int_0^u |\vec{q}(s)|_{g(z)} ds_{g(z)} = \left[\int_0^u |e^{-\alpha(z)+\alpha(y)}\vec{p}(s)| ds\right]_{g(z)}$$

$$= [e^{-\alpha(z)+\alpha(y)}L(p)]_{g(z)}. \tag{6.19}$$

This integral describes the length of the transported path as a sum of lengths of transported infinitesimal tangent vectors laid end to end as in Figure 6.3. This is the same path as that shown in Eq. (6.16).

Since parallel transports preserve numbers but not number values, the length of the parallel transport of $p_{g(y)}$ is the same as is the length of $p_{g(y)}$. But the length values are different. This is seen by

$$L(q)_{g(z)} = C_g(z,y)[L(p)]_{g(y)} = [e^{-\alpha(z)+\alpha(y)}]_{g(z)}[L(p)]_{g(z)}. \quad (6.20)$$

This shows that the length of the path $p_{g(y)}$ as a number in $\bar{R}_y^{g(y)}$ is the same number as is the length of the parallel transported path as a number in $\bar{R}_z^{g(z)}$. However, their length values differ by the exponential factor.

The paths $p_{g(y)}$ and $p_{g(z)}$, as images of the path p in M, are different. This is expressed by

$$C_g(z,y)\rho_{g(y)}(p) = C_g(z,y)p_{g(y)}$$
$$= [e^{-\alpha(z)+\alpha(y)}]_{g(z)}p_{g(z)} = [e^{-\alpha(z)+\alpha(y)}]_{g(z)}\rho_{g(z)}(p). \quad (6.21)$$

Figure 6.4 illustrates the relations between paths. It shows the paths $p_{g(y)}$ and $p_{g(z)}$ as chart images of p in M. Also shown is the parallel transport of $p_{g(y)}$ to $q_{g(z)}$.

A summary of the relations between paths and their values shows that $p_{y(y)}$ and $p_{g(z)}$ are different paths with the same path values. The values are provided by the local path environments, $\mathbb{S}_y^{g(y)}$ and $\mathbb{S}_z^{g(z)}$. The fact that the value of the path, $q_{g(z)}$, is different from p reflects the change of environment of the path $p_{g(y)}$ under parallel transport.

A good example of the relation between paths and other geometric quantities in spaces at different locations has the path p as a circle in M. The corresponding paths are circles in $\mathbb{S}_y^{g(y)}$ and $\mathbb{S}_z^{g(z)}$. For these paths, the initial and endpoints are the same. Let $\vec{u}_{g(y)}$ and $\vec{u}_{g(z)}$ be the radius vectors of the circles in $\mathbb{S}_y^{g(y)}$ and $\mathbb{S}_z^{g(z)}$. Then the circle circumferences are given by the path lengths as

$$L(p)_{g(y)} = [2\pi|\vec{u}|]_{g(y)} \quad \text{and} \quad L(p)_{g(z)} = [2\pi|\vec{u}|]_{g(z)}. \quad (6.22)$$

Here $|\vec{u}|_{g(y)} = (\vec{u}_{g(y)} \cdot_{g(y)} \vec{u}_{g(y)})^{1/2}$ is the length of $\vec{u}_{g(y)}$ as a number in $\bar{R}_y^{g(y)}$. Similar results hold for $\vec{u}_{g(z)}$. Parallel transport of the radius vector and circle from y to z gives the relations between the vector

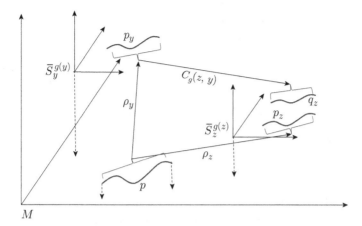

Figure 6.4. A schematic illustration of the relations between the values of paths p_y and q_z, and p_z. To save space, the path subscripts, $g(y)$ and $g(z)$, are shortened to just y and z. The two space images, $\mathbb{S}_y^{g(y)}$ and $\mathbb{S}_z^{g(z)}$, of M are shown along with their path contents. The chart maps of p in M to paths in $\mathbb{S}_y^{g(y)}$ and $\mathbb{S}_z^{g(z)}$ are shown as ρ_y and ρ_z. The compression of the path q_y relative to p_z illustrates the effect of scaling on path values. For this example, $g(y)/g(z) < 1$. The lengths of the dashed vertical arrows indicate the vertical components of the locations of the points, y, z, and two path points. The arrow heads give the horizontal locations of these points.

values of $\vec{u}_{g(y)}$ and $\vec{u}_{g(z)}$ and the circle values of $p_{g(y)}$ and $p_{g(z)}$. The relations are

$$\vec{u}_{g(y)} \to [e^{-\alpha(z)+\alpha(y)}\vec{u}]_{g(z)} \qquad (6.23)$$

and

$$L(p)_{g(y)} \to [e^{-\alpha(z)+\alpha(y)}L(p)]_{g(z)} = [e^{-\alpha(z)+\alpha(y)}2\pi|\vec{u}|]_{g(z)}. \qquad (6.24)$$

These two equations are for vectors in $\mathbb{S}_z^{g(z)}$ and numbers in $\bar{R}_z^{g(z)}$.

The understanding of the relationships between paths in $\mathbb{S}_y^{g(y)}$ and $\mathbb{S}_z^{g(z)}$ is helped by a representation of circles in the two images of M. This is done in Figure 6.5. To save on clutter, r represents $g(y)$ and s represents $g(z)$.

The figure shows the parallel transport of the radius vector and circle in $\mathbb{S}_y^{g(y)}$ to radius vectors and circles in $\mathbb{S}_z^{g(z)}$ for $r/s > 1$ and $r/s < 1$. Since transports preserve vectors, the visual length of the radius vector, $\vec{u}_r$, in the left-hand coordinate system is the same as

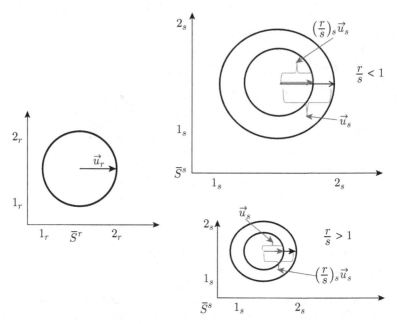

Figure 6.5. Representations in $\mathbb{S}_z^s$ of a circle in $\mathbb{S}_x^r$ for two values, $r/s < 1$ and $r/s > 1$. The radius vectors for each of the circles are shown. Coordinate systems indicate the relative scale. $\vec{u}_r$ and $\vec{u}_s$ are different vectors. But they have the same value, $\vec{u}$, in their respective images of M. The difference in length is shown by the expansion and contraction of the visual representations of $\vec{u}_s$ relative to $\vec{u}_r$. The fact that their length values are the same is shown by the expansion and contraction of the right-hand coordinate systems relative to the left-hand one $\mathbb{S}_y^r$. For ease in visualization, just two dimensions are shown.

the visual length of $(r/s)\vec{u}_s$ in both right-hand coordinate systems. The difference in their length values is shown by the expansion and contraction of the fiducial marks in the right-hand coordinate systems. The relationships between the sizes of the different circles in the coordinate systems show the same effects as does the relationships between the radius vectors.

6.3.2   *Paths as sections on the fiber bundle,* $\mathfrak{MF}^g$

So far, paths in M have been lifted as whole paths to local images of M at different locations of M. Relations between paths in images at different location have been described using parallel transport as connections to map paths from one image to another at a different location. The local image spaces are flat as is M.

The lifting of whole paths, as extended objects, to images in local spaces is not consistent with the approach taken in gauge theories. The field values as vectors belong to vector spaces collocated with the field values. The corresponding approach here begins with the representation of paths by a bundle of differential tangent vectors, $\{d\vec{p}(s) : a \leq s \leq b\}$. The path beginning and end locations are $p(a)$ and $p(b)$. The bundle is treated as a section on the fiber bundle with each differential tangent vector lifted to a differential vector in the collocated space.

Figure 6.3 shows the representation of a path by a bundle of tangent vectors. This is done by showing a path approximation by tangent vectors at closely spaced path points. Each path point has an associated small vector that is tangent to the path at the path point.

For each s, the image of the tangent vector $d\vec{p}(s)$ in M is the vector $[d\vec{p}(s)]_{g(p(s))}$ in the local space $\mathbb{S}_{p(s)}^{g(p(s))}$. This space is a component of the fiber, $F_{p(s)}^{g(p(s))}$, located at $p(s)$ in M. The relation between $[d\vec{p}(s)]_{g(p(s))}$ in $\mathbb{S}_{p(s)}^{g(p(s))}$ and $d\vec{p}(s)$ in M is given by

$$[d\vec{p}(s)]_{g(p(s))} = \rho_{p(s)}(d\vec{p}(s)). \tag{6.25}$$

Here $\rho_{p(s)}$ is the chart at $p(s)$ in M. The image tangent vector components are $[dp^j/ds]_{g(p(s))}$ with $j = 1 - 3$.

Properties of the path that require integration over the path or derivatives of the tangent vectors cannot be done with the setup described so far. The reason is that each image space, $\mathbb{S}_{p(s)}^{g(p(s))}$, contains just one tangent vector.

As integrals, path lengths are not defined because the integrands are numbers in local real number structures at the different locations, $p(s)$, in M. Addition between these quantities is not defined.

The methods used before will remedy this situation. One can create each numerical integrand, $[\|d\vec{p}(s)\|]_{g(p(s))}$, in the local space at $p(s)$ and then parallel transport the numerical integrand to a reference location, x, and then integrate. Or one can parallel transport the lifted tangent vectors to x, form the integrands, and then integrate.

The first method, parallel transport of the length integrand, gives

$$\mathcal{L}(p)_{g(x)} = \int C_g(x, g(p(s)))[|\vec{p}(s)|_{g(p(s))} ds]_{g(p(s))}$$

$$= [e^{-\alpha(x)}]_{g(x)} \int [e^{\alpha(p(s))}]_{g(x)}[|\vec{p}(s)|ds]_{g(x)}$$

$$= [e^{-\alpha(x)}]_{g(x)} \left[\int e^{\alpha(p(s))} \vec{p}(s)|ds \right]_{g(x)}. \tag{6.26}$$

The second method gives[3]

$$L(q_{g(x)}) = \int |\vec{q}_{g(x)}(s)| ds_{g(x)} = \int |C_g(x, p(s))\vec{p}(s)_{g(p(s))}| ds_{g(p(s))}$$

$$= [e^{-\alpha(x)}]_{g(x)} \left[\int e^{\alpha(p(s))} |\vec{p}(s)| ds \right]_{g(x)}$$

$$= \left[\int |\vec{q}(s)| ds \right]_{g(x)} = L(q)_{g(x)}. \tag{6.27}$$

Comparison of this result with Eq. (6.26) shows that

$$\mathcal{L}(p)_{g(x)} = L(q)_{g(x)}. \tag{6.28}$$

The tangent vectors $\vec{q}(s)_{g(x)}$ are the parallel transports of $\vec{p}(s)_{g(p(s))}$ to x as in

$$\vec{q}(s)_{g(x)} = C_g(x, p(s))[\vec{p}(s)]_{g(p(s))} = [e^{-\alpha(x)+\alpha(p(s))} \vec{p}(s)]_{g(x)}. \tag{6.29}$$

Here $\vec{q}(s)_{g(x)}$ is the same vector in $\mathbb{S}_x^{g(x)}$ as $\vec{p}(s)_{g(p(s))}$ is in $\mathbb{S}_{p(s)}^{g(p(s))}$. Their values differ by the exponential factor.

The relation between the tangent vectors, $\vec{q}_{g(x)}(s)$ and $\vec{p}_{g(x)}(s)$, is shown in Figure 6.6. To keep the figure clear, the relationship is shown in two-dimensional Euclidean space. The third dimension represents scaling or vector value levels of the local spaces. The figure shows that, for each s, the tangent vector, $\vec{p}_{g(p(s))}(s)$, lifted from the tangent bundle representation of the path, p in M, is parallel transported to become $\vec{q}(s)_{g(x)}$. For each s, the vector, $\vec{p}(s)_{g(p(s))}$, is located at $p(s)_{g(p(s))} = P_{g(p(s))}(s)$ in $\mathbb{S}_{p(s)}^{g(p(s))}$.

---

[3]This is an example of the one case in which parallel transport commutes with multiplication. Commutation holds if the integrand is the nth root of n parallel transported collocated factors. Here $n = 2$.

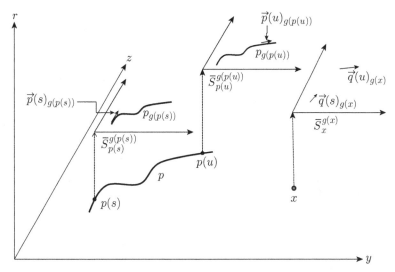

Figure 6.6. Representation of the parallel transport of the tangent vector of p in each of $\mathbb{S}_{p(s)}^{g(p(s))}$ and $\mathbb{S}_{p(u)}^{g(p(u))}$ to tangent vectors of $q_{g(x)}$ in $\mathbb{S}_x^{g(x)}$. The two coordinate dimensions, y and z, of M are shown. The third dimension, r, is a scaling or vector value level dimension. The heights of the image spaces above the y–z plane, shown by the dashed arrows, indicates the relative magnitude of the scaling factors, $g(p(s))$, $g(p(u))$, and $g(x)$. This is also indicated by showing the tangent vector, $\vec{q}(s)_{g(x)}$, shorter than $\vec{p}(s)_{g(p(s))}$ and $\vec{q}(u)_{g(x)}$ longer than $\vec{p}(u)_{g(p(u))}$.

Figure 6.6 also shows the effect of space-dependent value factors on two tangent vectors of p. It is useful to see how this affects the change in the shape of path p as shown by the path q. This is done in Figure 6.7 in a local three-dimensional space $\mathbb{S}_x^{g(x)}$. The path $p_{g(x)}$ can be considered as a lift of a path p in M, as a whole to a path in $\mathbb{S}_x^{g(x)}$.

The figure illustrates the difference between the paths $q_{g(x)}$ and $p_{g(x)}$ for a simple case where $g(p(s))/g(x) = 3/2$ for $0 \leq s \leq 0.5$ and $g(p(s))/g(x) = 2/3$ for $0.5 \leq s \leq 1$. This shows that the path $q_{g(x)}$ is a distortion of the path $p_{g(x)}$ where the distortion is given by the variation of $\alpha(p(s))$ with s.

The path length of Eq. (6.27) has the property that parallel transport of $L(q_{g(x)})$ to another location, z, gives the same result as does direct localization to z in the first place. That is,

$$L(q_{g(z)}) = C_g(z, x)L(q_{g(x)}). \tag{6.30}$$

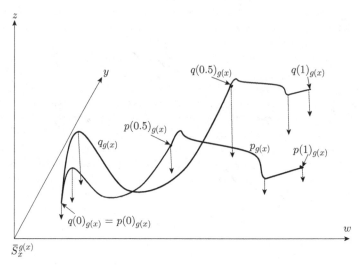

Figure 6.7. Relation between paths $p_{g(x)}$ and $q_{g(x)}$ in the vector space $\mathbb{S}_x^{g(x)}$. The path $q_{g(x)}$ is obtained from that for $p_{g(x)}$ by setting $g(p(s))/g(x) \approx 3/2$ for $0 \leq s < 0.5_{g(x)}$ and $g(p(s)/g(x) \approx 2/3$ for $0.5 \leq s < 1_{g(x)}$. The first half of $q_{g(x)}$ expands the first half of $p_{g(x)}$ by a factor of $\approx 3/2$. The second half of $q_{g(x)}$ contracts the second half of $p_{g(x)}$ by a factor of $\approx 2/3$. The vertical dashed arrows to the w, y plane attempt to give a three-dimensional representation to the curves.

Here $L(q_{g(z)})$ is given by Eq. (6.27) with z replacing x everywhere. This representation of the path length also has the property that the effects of the value field disappear if the value field, α, is a constant. The effects are also very small if α varies little over the path sites, x and z.

Examples help illustrate the effect of representation of paths as sections on the fiber bundle, $\mathfrak{MF}^g$. The simplest example is a path as a straight line. In the usual global mathematics, this means that the tangent vectors, $\vec{p}(s)$ and $\vec{p}(t)$, of the path p at locations, $p(s)$ and $p(t)$, are parallel to one another. They should also have the same length.

In the arena of local mathematics, comparison of the two vectors to determine if they are parallel or not, or have the same length, requires parallel transport of the vectors to a common reference point for comparison. Explicitly, one has

$$[\vec{q}(s)]_{g(z)} = C_g(z, g(p(s)))[\vec{p}(s)]_{g(p(s))} = [e^{-\alpha(z)+\alpha(p(s))}\vec{p}(s)]_{g(z)}$$
$$(6.31)$$

and

$$[\vec{q}(t)]_{g(z)} = C_g(z, g(p(t)))[\vec{p}(t)]_{g(p(t))} = [e^{-\alpha(z)+\alpha(p(t))}\vec{p}(t)]_{g(z)}.$$
(6.32)

Since the vectors $\vec{q}(s)_{g(z)}$ and $\vec{q}(t)_{g(z)}$ are in the same space, $\mathbb{S}_z^{g(z)}$, they can be compared.

Equations (6.31) and (6.32) show that parallel transport of vectors multiplies the vectors by scalars. One knows that multiplying vectors by scalars does not affect whether they are parallel or not. It follows that, the vector, $\vec{q}(s)_{g(z)}$, is parallel to $\vec{q}(t)_{g(z)}$ if and only if $\vec{p}(s)_{g(z)}$ is parallel to $\vec{p}(t)_{g(z)}$. This follows from the fact that for each s and t

$$\frac{\vec{q}(s)_{g(z)} \cdot \vec{q}(t)_{g(z)}}{|\vec{q}(s)_{g(z)}||\vec{q}(t)_{g(z)}|} = 1_{g(z)} \qquad \Leftrightarrow \qquad \frac{\vec{p}(s)_{g(z)} \cdot \vec{p}(t)_{g(z)}}{|\vec{p}(s)_{g(z)}||\vec{p}(t)_{g(z)}|} = 1_{g(z)}. \quad (6.33)$$

Another more complicated example is the determination of whether a path, as a bundle of tangent vectors each in separate vector spaces, is a circle. To keep the discussion simple, the description will be restricted to M as a two-dimensional Euclidean space.

The criterion used to determine if a path is a circle is that the curvature at all path points is a constant. If it is a constant, then the radius of the circle is the inverse of the curvature [70].

A good aspect of the curvature condition is that curvature is local in that it must be determined at each path point. Comparison of curvatures will require parallel transport to a common location for comparison.

Let p be a closed path parameterized by s with endpoints identified. The path is represented here by a bundle of tangent vectors at each path point. The lifted tangent vector at $p(s)$ is represented in terms of the x and y components as

$$[\vec{p}(s)]_{g(p(s))} = \left[\frac{dp_x}{ds}\hat{x} + \frac{dp_y}{ds}\hat{y}\right]_{g(p(s))} = p'_x(s)\hat{x} + p'_y(s)\hat{y}. \quad (6.34)$$

The curvature of the path at $p(s)$ is defined by the scalar quantity [70],

$$\kappa(p(s))_{g(p(s))} = \left[\frac{|p'_x(s)p''_y(s) - p''_x(s)p'_y(s)|}{(p'_x(s)^2 + p'_y(s)^2)^{3/2}}\right]_{g(p(s))}. \quad (6.35)$$

The primes on $p_x(s)$ and $p_y(s)$ denote the x and y components of the first and second derivatives of the path at $p(s)$. Here $\kappa(p(s))_{g(p(s))}$ is a numerical quantity in $\bar{R}_{p(s)}^{g(p(s))}$ with value $\kappa(p(s))$.

There are two problems with this definition. One is the question of whether the curvature, as written, is defined. The other is the fact that the requirement that $\kappa(p(s))_{g(p(s))}$ be the same for all values of s requires direct comparison of the quantities, $\kappa(p(s))_{g(p(s))}$, for different values of s. Both these problems stem from the fact that, for each value of s, the path second derivative is not defined within $\mathbb{S}_{p(s)}^{g(p(s))}$ and comparison of curvatures at different path points is not possible because the curvatures are in different real number structures.

There are two ways to solve these problems. One is to alter the definitions of the curvatures so that they are well defined and then to parallel transport the curvatures to a common location for comparison. The other way is to parallel transport the path tangent vectors to a common location and then compare the curvatures determined from the path tangent vectors all at the same location. As will be seen, the two methods give different results.[4]

The first method begins by noting that the definition of the curvature in Eq. (6.35) is not defined. The problem is that the second derivatives, defined by

$$p_x''(s) = \frac{p_x'(s+ds)_{g(p_x(s+ds))} - p_x'(s)_{g(p(s))}}{ds}, \qquad (6.36)$$

are not defined because the subtraction in the numerator is between vectors in different vector spaces. This is fixed by parallel transport of the left-hand term in the numerator to the vector space containing the right-hand term. For the x component of the path, one obtains

$$[p_x''(s)]_{g(p(s))}$$

$$= \frac{C_g(p(s), p_x(s+ds))p_x'(s+ds)_{g(p(s+ds))} - p_x'(s)_{g(p(s))}}{ds}$$

$$= \left[\frac{e^{-\alpha(p(s))+\alpha(p_x(s+ds))}p_x'(s+ds) - p_x'(s)}{ds} \right]_{g(p(s))}. \qquad (6.37)$$

---

[4]This difference will be accounted for in the following chapter. It will be seen that the variation in the α field, which is the cause of the difference, is too small to have a local effect.

Taylor expansion of the exponent term, $\alpha(p_x(s+ds))$, to first order in small quantities gives the exponential factor $e^{ds A_x(p(s))p'_x(s)}$. Here $A_x(p(s)) = d\alpha(p_x(s))/dp_x(s)$ is the gradient component of $\alpha(p(s))$ in the direction $\hat{x}$. Expansion of the exponential to first order gives

$$p''_x(s) = (\partial_{x,s} + A_x(p(s))p'_x(s))p'_x(s) = D_{x,s}p'_x(s). \tag{6.38}$$

The equation is defined because the terms are all quantities in the fiber at $p(s)$.

Carrying out a similar parallel transport for the y path component gives

$$p''_y(s) = (\partial_{y,s} + A_y(p(s))p'_y(s))p'_y(s) = D_{y,s}p'_y(s). \tag{6.39}$$

The resulting curvature at $p(s)$ becomes

$$\kappa(p(s))_{g(p(s))} = \left[\frac{|p'_x(s)D_{y,s}p'_y(s) - D_{x,s}p'_x(s)p'_y(s)|}{(p'_x(s)^2 + p'_y(s)^2)^{3/2}} \right]_{g(p(s))}. \tag{6.40}$$

The definitions of these curvatures for different path points are all valid. However, the curvatures for different values of s cannot be compared because they are all numbers in number structures, $\bar{R}^{g(p(s))}_{p(s)}$, at different locations.

This is fixed by parallel transport of the curvatures to a common location, z. The transported number valued quantities are given by

$$K(p(s)) = e^{-\alpha(z)+\alpha(p(s))} \frac{|p'_x(s)D_{y,s}p'_y(s) - D_{x,s}p'_x(s)p'_y(s)|}{(p'_x(s)^2 + p'_y(s)^2)^{3/2}}. \tag{6.41}$$

The curvatures for different path locations can now be compared as they are all values of numerical quantities in the fiber at z.

The other way to solve both these problems is to parallel transport the factors in the definition of the curvature at each path location separately to a reference location, z, and then form the curvature from the transported factors. The effect of transport on the factors of Eq. (6.40) gives

$$C_g(z, p(s))p_x'(s) = e^{-\alpha(z)+\alpha(p(s))}p_x'(s) = q_x'(s),$$
$$C_g(z, p(s))p_y'(s) = e^{-\alpha(z)+\alpha(p(s))}p_y'(s) = q_y'(s),$$
$$C_g(z, p(s))D_{x,s}p_x'(s) = e^{-\alpha(z)+\alpha(p(s))}D_{x,s}p_x'(s) = q_x''(s),$$
$$C_g(z, p(s))D_{y,s}p_y'(s) = e^{-\alpha(z)+\alpha(p(s))}D_{y,s}p_y'(s) = q_y''(s).$$

(6.42)

Use of these factors to determine the curvature as the value of a number at z gives

$$\kappa(p(s)) = e^{+\alpha(z)-\alpha(p(s))}\frac{|p_x'(s)D_{y,s}p_y'(s) - D_{x,s}p_x'(s)p_y'(s)|}{(p_x'(s)^2 + p_y'(s)^2)^{3/2}}. \quad (6.43)$$

The corresponding number with this value in $\bar{R}_z^{g(z)}$ is obtained by appending the $g(z)$ subscript to the terms in this equation.

The value of $\kappa(p(s))$ in Eq. (6.43) is different from that of Eq. (6.40) in that the exponential factor in Eq. (6.40) is the inverse of that in the above equation. This difference is an example of the fact that, in the presence of scaling or valuation, parallel transport of a term consisting of products and inverses of factors does not commute with the process of parallel transport of the individual factors and then forming the curvature term at the reference location.

One sees from these results that $K(p(s))$ is independent of s if and only if $\kappa(p(s))$ is independent of s. It follows that both curvatures describe circular paths with radii $K(p(s))^{-1}$ and $\kappa(p(s))^{-1}$. The circles are of different sizes because the radii are different.

One cannot conclude from the constancy of these curvatures that the localization of the path p to a path in $\mathbb{S}_z^{g(z)}$ is a circle. This is due to the dependence of $\alpha(p(s))$ on s. This is the case even if the path, p, on M is a circle. However, it does follow that the path q is a circle. This follows from the use of Eq. (6.43) to write

$$\kappa(q(s)) = \frac{|q_x'(s)q_y''(s) - q_x''(s)q_y'(s)|}{(q_x'(s)^2 + q_y'(s)^2)^{3/2}}. \quad (6.44)$$

Since the curvature of Eq. (6.44) is the same as that of Eq. (6.43), $\kappa(p(s)) = \kappa(q(s))$, the circular path, q, has the same radius as does the circle with radius $1/\kappa(p(s))$.

So far two different methods of determining if a path is a circle have been discussed. These methods give different results. This is shown by the fact that the exponential factor that scales the curvature in Eq. (6.41) is the inverse of the exponential factor scaling the curvature in Eq. (6.43). Which method is chosen depends on the context in which the curvature is defined. In general, it would seem preferable to determine the curvature of a localized path, $q_{g(z)}$, in a fixed local coordinate systems, $\mathbb{S}_z^{g(z)}$, rather than separate parallel transport of the curvature at each path point, $p(s)$, in $\mathbb{S}_{p(s)}^{g(p(s))}$ to a reference location.

In any case, as will be seen, the variation in the value field α will be much too small to be seen experimentally. It follows that within experimental error, the two methods are consistent and that the path, p, can be identified with q. Caution, this does not mean that variations in the α field are small everywhere or have no effect. It does mean that variations in α are small locally. This will be clarified more later on.

6.3.2.1 *Parallel transports and conformal transformations*

The results shown in Figure 4.2 show that parallel transport of geometric quantities includes expansion or contraction of vector lengths. This suggests that parallel transport may be similar to conformal transformations. These transformations change vector lengths but not angles between vectors [71].

In spite of this apparent similarity, parallel transports as used here are quite different from conformal transformations. The mathematical background for conformal transformations consists of the usual global mathematics. The mathematical background for parallel transports consists of local mathematics with the values of numerical physical quantities depending on the locations of the quantities. In addition, angles and their sines and cosines are changed under parallel transport. This is not the case for conformal transports.

The apparent similarity stems from the following observation. Let $\vec{w}_{g(y)}$ and $\vec{v}_{g(y)}$ be two vectors in $\mathbb{S}_y^{g(y)}$. The cosine of the angle between the vectors is defined by

$$[\cos(\theta)]_{g(y)} = \frac{\vec{w}_{g(y)} \cdot \vec{v}_{g(y)}}{|\vec{w}_{g(y)}||\vec{v}_{g(y)}|}. \tag{6.45}$$

Here $|\vec{w}_{g(y)}|$ denotes the length of the vector $\vec{w}_{g(y)}$ as a number in $\bar{R}_y^{g(y)}$. The scalar product and implied multiplications and division are all operations in $\mathbb{S}_y^{g(y)}$ and $\bar{R}_y^{g(y)}$.

Parallel transport of the vectors to $\mathbb{S}_x^{g(x)}$ at location x gives the vectors $[(g(y)/g(x))\vec{w}]_{g(x)}$ and $[(g(y)/g(x))\vec{v}]_{g(x)}$. The cosine of the angle between the two vectors is given by

$$[\cos(\theta)]_{g(x)} = \frac{l\vec{w}_{g(x)} \cdot l\vec{v}_{g(x)}}{|l\vec{w}_{g(x)}||l\vec{v}_{g(x)}|} = \frac{\vec{w}_{g(y)} \cdot \vec{v}_{g(y)}}{|\vec{w}_{g(y)}||\vec{v}_{g(y)}|}, \qquad (6.46)$$

where $l = e^{-\alpha(x)+\alpha(y)}$. Conformal transformation of the vectors $\vec{w}_{g(x)}$ and $\vec{v}_{g(x)}$ by the number l also gives the middle term of Eq. (6.46) for the cosine of the angle between the vectors. Comparison of Eqs. (6.45) and (6.46) shows that the numbers $[\cos(\theta)]_{g(y)}$ and $[\cos(\theta)]_{g(x)}$ are different numbers that have the same value.

6.3.3 *Trigonometric relations*

Unlike the case for conformal transformations, both angles and trigonometric functions of angles are changed under parallel transport. Trigonometric relations remain unchanged.

Power series descriptions of $\cos(\theta)$ and $\sin(\theta)$ are a good way to determine the effect of parallel transports on these functions. At location y, the cosine expansion, using numbers in $\bar{R}_y^{g(y)}$, is

$$\cos_{g(y)}(\theta_{g(y)}) = \sum_{n=0}^{\infty} \frac{(\theta_{g(y)})^{2n}}{[2n]_{g(y)}!}. \qquad (6.47)$$

All the factors in this equation, and the multiplication and division operations, are numbers and operations in $\bar{R}_y^{g(y)}$.

One can use the fact that

$$[\cos(\theta)]_{g(y)} = \left[\sum_{n=0}^{\infty} \frac{\theta^{2n}}{2n!}\right]_{g(y)} = \sum_{n=0}^{\infty} \frac{(\theta_{g(y)})^{2n}}{[2n]_{g(y)}!} \qquad (6.48)$$

to obtain

$$[\cos(\theta)]_{g(y)} = \cos_{g(y)}(\theta_{g(y)}). \qquad (6.49)$$

The subscript $g(y)$ on $\cos_{g(y)}$ emphasizes the fact that the factors and operations appearing in each term of the power series expansion are elements of $\bar{R}_y^{g(y)}$.

Parallel transport on both sides of Eq. (6.48) to location z gives

$$[e^{-\alpha(z)+\alpha(y)}]_{g(z)}[\cos(\theta)]_{g(z)} = [e^{-\alpha(z)+\alpha(y)}]_{g(z)}\left[\sum_0^\infty \frac{\theta^{2n}}{2n!}\right]_{g(z)}.$$
(6.50)

Cancellation of the exponential factors on both sides of the equation gives

$$[\cos(\theta)]_{g(z)} = \left[\sum_{n=0}^\infty \frac{\theta^{2n}}{2n!}\right]_{g(z)}.$$
(6.51)

This is an example of the invariance of equations under parallel transport.

Parallel transport of Eq. (6.47) from y to z provides an illustration of the transport of individual quantities and associated operations. One has

$$C_g(z,y)\cos_{g(y)}(\theta_{g(y)}) = C_g(z,y)\sum_{n=0}^\infty \frac{(\theta_{g(y)})^{2n}}{[2n]_{g(y)}!}$$

$$= \sum_{n=0}^\infty r_{g(z)}\frac{[-r]_{g(z)}[1/r\times]_{g(z)}[r\theta]_{g(z)}[1/r)\times]_{g(z)}([1/r\times]_{g(z)}([r\theta]_{g(z)}))^{2n-1}}{[r2n])_{g(z)}[1/r\times]_{g(z)}),\ldots,[r1]_{g(z)}}$$

$$= r_{g(z)}\sum_{n=0}^\infty \frac{(\theta_{g(z)})^{2n})}{([2n]_{g(z)}))!} = r_{g(z)}\cos_{g(z)}(\theta_{g(z)}) = [r\cos(\theta)]_{g(z)}.$$
(6.52)

In this complex equation, notation is saved by setting $r = e^{-\alpha(z)+\alpha(y)}$. The r^{-1} factors arise from the transport of $\times_{g(y)}$ to the scaled multiplication operation in $\bar{R}_z^{g(z)}$. The r factors in the numerator of the fraction cancel to a single r factor. This is canceled by the single r factor in the denominator. The r factor in front of the fraction accounts for the parallel transport of the division operation.

In Eq. (6.52),

$$[r\theta]_{g(z)} = [e^{-\alpha(z)+\alpha(y)}]_{g(z)}[\theta]_{g(z)} = [e^{-\alpha(z)+\alpha(y)}\theta]_{g(z)}$$

is the same angle as is $\theta_{g(y)}$. But the angle value differs by the factor r. The angle, $\theta_{g(z)}$, has the same angle value as does $\theta_{g(y)}$. But $\theta_{g(z)}$ is a different angle from $\theta_{g(y)}$.

Equation (6.52) shows that parallel transport of $[\cos(\theta)]_{g(y)}$ to z multiplies the cosine value by r. The angle value and cosine value remain the same. But the cosine value at z of the parallel transport of θ gives a different result:

$$\cos_{g(z)}(r\theta_{g(z)}) = [\cos(r\theta)]_{g(z)}. \tag{6.53}$$

Here, $\cos_{g(z)}(r\theta_{g(z)})$ is a different number with a different value, $\cos(r\theta)$, than is the cosine value, $\cos(\theta)$, of $[r\cos(\theta)]_{g(z)}$.

The differences between the two methods of parallel transport, parallel transport of the cosine of an angle and the cosine of a parallel transported angle, are illustrated in Figure 6.8. As expected, the two methods give different results.

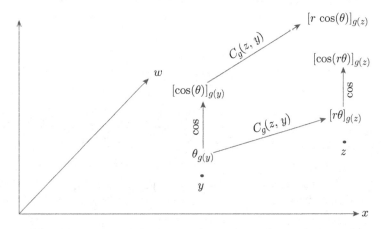

Figure 6.8. Illustration of the two methods of parallel transport: transport of the cosine of an angle and the cosine of a parallel transported angle. Both are from point y to point z. The points are shown in the two-dimensional w, x plane. The third axis is present to distinguish the values of the three different cosine numbers. The vertical arrows indicate the map $\theta \to \cos(\theta)$. The figure is shown as an open diagram. This shows the noncommutativity of parallel transport with multiplication and division operations.

Trigonometric relations are preserved for both transport of $[\sin(\theta)]_{g(y)}$ and $[\cos(\theta)]_{g(y)}$ from y to z. They are also preserved for the sine and cosine of the transport of $\theta_{g(y)}$ to z.

For example, the relation

$$[\sin(\theta)]^2_{g(y)} + [\cos(\theta)]^2_{g(y)} = 1_{g(y)} \tag{6.54}$$

is a valid relation for the numbers, $[\sin(\theta)]^2_{g(y)}$ and $[\cos(\theta)]^2_{g(y)}$, in $\bar{R}^{g(y)}_y$ at y. Parallel transport of the terms in this equation to z gives

$$r_{g(z)}[\sin(\theta)]^2_{g(z)} + r_{g(z)}[\cos(\theta)]^2_{g(z)} = r_{g(z)}1_{g(z)}. \tag{6.55}$$

Here $r = e^{\alpha(y)-\alpha)z)}$. Cancellation of the common $r_{g(z)}$ factor gives

$$[\sin(\theta)]^2_{g(z)} + [\cos(\theta)]^2_{g(z)} = 1_{g(z)}. \tag{6.56}$$

This is another example of preservation of equations under parallel transport. Also $[\sin(\theta)]_{g(z)}$ and $[\cos(\theta)]_{g(z)}$ have the same number values as do $[\sin(\theta)]_{g(y)}$ and $[\cos(\theta)]_{g(y)}$.

The same relation for the transported angle, $[r\theta]_{g(z)}$, at z is

$$\sin^2([r\theta]_{g(z)}) + \cos^2([r\theta]_{g(z)})^2 = 1_{g(z)}. \tag{6.57}$$

Here $[r\theta]_{g(z)}$ is the same angle at z as is $\theta_{g(y)}$ at y. As can be seen, the angle values are different.

6.3.4 *Effect of α field on the sizes of local spaces*

The presence of the meaning field, α, the "no information at a distance" principle, and local mathematics have an effect on the relative sizes of the local spaces. This effect is a result of the properties of the connection that implements parallel transports of structures between two local spaces.

Let $\mathbb{S}^{g(y)}_y$ and $\mathbb{S}^{g(z)}_z$ be local spaces at locations y and z in M. Let $[d\vec{w}]_{g(y)}$ and $[|d\vec{w}|]_{g(y)}$ be a small vector and vector length at an arbitrary location, w, in the local space at y. Parallel transport of $[d\vec{w}]_{g(y)}$ to the local space at z gives the vector, $[e^{-\alpha(z)+\alpha(y)}d\vec{w}]_{g(z)}$.

The length of this vector is given by

$$[[e^{-\alpha(z)+\alpha(y)}\,d\vec{w}]]_{g(z)} = [(e^{-2\alpha(z)+2\alpha(y)}\delta_{j,k}dw^j\,dw^k)^{1/2}]_{g(z)}. \quad (6.58)$$

Parallel transport of the length, $[[d\vec{w}]]_{g(y)}$, to the local space and real number structure at z gives the length

$$[e^{-\alpha(z)+\alpha(y)}|d\vec{w}]]_{g(y)} = [[e^{-\alpha(z)+\alpha(z)}d\vec{w}]]_{g(z)}$$
$$= [(e^{-2\alpha(z)+2\alpha(y)}\delta_{j,k}dw^j\,dw^k)^{1/2}]_{g(z)}. \quad (6.59)$$

Comparison of the results in Eqs. (6.58) and (6.59) shows that the length of the transported vector is equal to the transported length of the vector.

The relative sizes between spaces are determined by the relations between the metric tensors for the spaces. Since the local spaces at y and z are Euclidean, the metric tensors for these spaces are $[\delta_{j,k}]_{g(y)}$ and $[\delta_{j,k}]_{g(z)}$.

The relation between the sizes of the two local spaces requires comparing the values of the metric tensors for the two spaces. They cannot be directly compared because they are numbers in real number structures at different locations. Comparison requires parallel transport of the number at z to y or transport of the number at y to z. Here parallel transport to z is considered.

Equations (6.58) and (6.59) give the results of the parallel transport. They show the metric tensor at y as seen from z. The sign of the exponent gives the size of the space at y as seen from the space at z. If the exponent is positive, the space at y is an expansion of the space at z. If the exponent is negative, the space at y is a contraction of the space at z. If the exponent is zero, the two spaces have the same sizes.

This relation between the relative sizes of the spaces does not change the fact that, with one exception, the local spaces are all Euclidean. The exception is the space chosen as a reference space. To see this, let y be any point in M and $d\vec{y}$ be a differential vector at y in M.

Lift of the vector $d\vec{y}$ to the vector, $[d\vec{y}]_{g(y)}$, at the point $y_{g(y)}$ in the local space, $\mathbb{S}_y^{g(y)}$, followed by parallel transport to the point, $y_{g(z)}$, in $\mathbb{S}_z^{g(z)}$ gives the vector $[e^{-\alpha(z)+\alpha(y)}d\vec{y}]_{g(z)}$ in the local space at z. The length of this vector is shown by

$$[e^{-\alpha(z)+\alpha(y)}|d\vec{y}|]_{g(z)} = \left[\left(e^{-2\alpha(z)+2\alpha(y)}\sum_{j,k}\delta_{j,k}\partial^j y\partial^k y\right)^{1/2}\right]_{g(z)}.$$

(6.60)

The same result is obtained by lifting the length $|d\vec{y}|$ of the vector $d\vec{y}$ in M to the length $|d\vec{y}|_{g(y)}$ in the local space at y. Parallel transport of this length to the local space at z gives[5] $[e^{-\alpha(z)+\alpha(y)}|d\vec{y}|]_{g(z)}$.

Let w and $d\vec{w}$ be a location and differential vector at w in M. Lift of the vector to the location, $w_{g(w)}$, in the space at w, followed by parallel transport to the space at z gives the vector, $[e^{-\alpha(z)+\alpha(w)}d\vec{w}]_{g(z)}$, at location, $w_{g(z)}$, in the space at z. If the exponent is positive or negative, the vector $[\vec{d(w)}]_{g(w)}$ in the space at w is respectively longer or shorter than the vector $[d\vec{w}]_{g(z)}$ at the point $w_{g(z)}$ in the reference space at z.

This can be used to divide $\mathbb{S}_z^{g(z)}$ into regions of points, $w_{g(z)}$, in which the value of the exponent of, $[e^{-\alpha(z)+\alpha(w)}]_{g(z)}$, is either positive, negative, or zero. The same divisions apply to the lengths, $[|d\vec{w}|]_{g(w)}$ and $[|d\vec{w}|]_{g(z)}$, in the spaces at w and z. The three regions defined for vectors being longer or shorter or the same correspond to those for vector lengths being larger, smaller, or the same as the vector lengths in the space at z.

Figure 6.9 is an illustration of the regions in the local space at the reference location, z. Five regions are shown. The regions marked > 0 or < 0 are areas in which the local spaces at all points in a region are respective expansions or contractions of the space at z. For each point, $y_{g(z)}$ in a region, the local space at location, y, in M is an expansion or contraction of the space at z if the exponent $-\alpha(z) + \alpha(y)$ is positive or negative. Most of the space labeled 0 contains the locations of spaces that are the same size as the local

---

[5]This is an example of the case where lack of commutativity of multiplication and parallel transport has no effect. The square root of a product of two transported numbers is the same as the transport of the square root of two numbers.

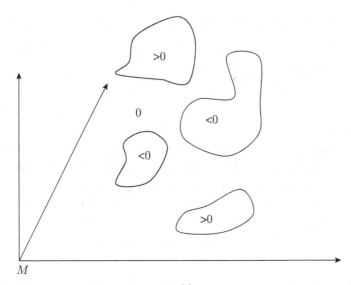

Figure 6.9. Example of regions in $\mathbb{S}_z^{g(z)}$ for which the transport exponent is greater than, less than, or equal to 0. These are regions in which the local spaces at each region point are respective expansions, contractions, or the same size as the space at z.

space at z. The relation between points in a region of the space at z and spaces at corresponding locations in M that are different from z is based on the fact that these spaces are Euclidean spaces. Metric tensors for each of these spaces are the same for all locations within each space. This is why the relation of points in the space at z is to spaces at other locations rather than to regions or points in the spaces.

Figure 6.10 illustrates the relation between vectors in local spaces at y and w to vectors in the reference space at z.

From Eq. (6.60), one sees that the metric tensor at the point $y_{g(z)}$ in the local space $\mathbb{S}_z^{g(z)}$ is

$$[e^{-2\alpha(z)+2\alpha(y)}\delta_{j,k}]_{g(z)}. \tag{6.61}$$

$y_{g(z)}$ is any point in $\mathbb{S}_z^{g(z)}$. This shows that the metric tensor for the reference space at z is different for different locations in the space. The difference is determined by the spatial variation of the α field.

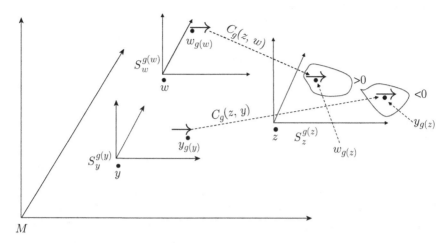

Figure 6.10. Representation of the relation between the vectors (shown as arrows) in the local spaces at y and w and the corresponding vectors in the reference space at z. The dashed arrows from vectors in the spaces at y and w labeled with connections indicate parallel transports of the vectors in the spaces at y and w to the space at z. The relative lengths of the vectors in the spaces at y and w to that of the vector in the space at z indicate whether the connection is an expansion or a contraction. The figure shows the space at w as an expansion and the space at y as a contraction relative to the space at z. The corresponding exponents are >0 and <0.

The form of the metric tensor in $\mathbb{S}_z^{g(z)}$ shows that the space is distorted. This is shown by the presence of $\alpha(y)$ in the metric tensor $[e^{-2\alpha(z)+2\alpha(y)}\delta_{j,k}]_{g(z)}$ for each point $y_{g(z)}$ in $\mathbb{S}_z^{g(z)}$.

The dependence of the metric tensor on locations in the reference space at z shows that the geometry of the local reference space at z is not Euclidean. The space is no longer flat. This is illustrated by the presence of the exponential factor $e^{\alpha(p(s))}$ in the integrand for the path length integral in the local space at z. This is given by

$$\left[e^{-\alpha(z)}\int_0^t (e^{2\alpha(p(s))}\delta_{j,k}dp^j(s)dp^k(s))^{1/2}\right]_{g(z)}. \tag{6.62}$$

Distances between points in $\mathbb{S}_z^{g(z)}$ are determined by finding the path p in M that minimizes the path length integral. The distance is determined by the Euler–Lagrange equations for the length integrand of Eq. (6.62). These equations are given by

$$\frac{d}{ds}\left(\frac{dp^j(s))/ds}{|\vec{p}(s)|}\right) = A_j(p(s))|\vec{p}(s)| - (\vec{A}(p(s)) \cdot \vec{p}(s))\frac{dp^j(s)/ds}{|\vec{p}(s)|}.$$

(6.63)

There are three of these Euler–Lagrange equations, one for each j. Here $|\vec{p}(s)| = (\delta_{j,k}dp^j(s)dp^k(s))^{1/2}$.

The equations show how the gradient of the α field affects the geometry of the reference space at z. The amount of deviation from flatness is determined by the strength of the gradient field. The Euler–Lagrange equations show that if the gradient field was 0 everywhere, the equations would be those for straight lines as distances between points. The geometry would be that for a flat space.

The Euler–Lagrange equations are valid for the space serving as a reference space, as is the local space at z. They do not hold for spaces at locations different from z. For these flat spaces, the Euler–Lagrange equations can be obtained from Eq. (6.63) by setting $\vec{A} = 0$ for all locations in the spaces.

6.4   Euclidean space and time

So far the discussion has been limited to three-dimensional local spaces located on a three-dimensional Euclidean space. Time has played no role. Since time plays an important role in physics and geometry, this defect should be remedied.[6]

The fiber bundle representation now becomes

$$\mathfrak{MTF}^g = M \times T, F, \pi_g, M \times T.$$

(6.64)

The fiber is defined by

$$F = \cup_r \cup_S S^r.$$

(6.65)

As before, the union over S is over all mathematical system types that include scalars in their description. This includes at least real

---

[6]This expansion was used earlier as a background for the description of quantum dynamics.

and complex number structures, $\bar{R}^r, \bar{C}^r$, vector spaces, $\bar{V}^r$, and a Euclidean three-dimensional space, $\mathbb{S}^r$. The union over r is over all real scaling factors.

The inverse, π_g^{-1}, of the projection operator is a map from $M \times T$ to a subset of $M \times T, F$. It is defined by

$$\pi_g^{-1}(y,s) = F_{y,s}^{g(y,s)} = \cup_S S_{y,s}^{g(y,s)}. \tag{6.66}$$

Space and time locations in $M \times T$ are denoted by y and s. To account for the inclusion of time, the g function's dependence on space only is expanded to include a dependence on time.

A path in $M \times T$ can be represented by a collection of path point pairs, $p(s), s$. The time is s and the space location of the path at time s is given by $p(s)$. There is no restriction on the path slope. It can be positive, negative, or 0.

The path p is also equivalent to a collection of differential tangent vectors $d\vec{p}(s)$ at each path point. Representation of the collection as a section on the fiber bundle results in the lift of each vector, $d\vec{p}(s)$, in the collection to a vector, $[d\vec{p}(s)]_{g(p(s),s)}$, in the local space, $\mathbb{S}_{p(s),s}^{g(p(s),s)}$, collocated with the vector.

The differential lengths, $|d\vec{p}(s)|$, are also lifted to numbers, $|d\vec{p}(s)|_{p(s),s}$, in a real number structure, $\bar{R}_{p(s),s}^{g(p(s),s)}$, at the same location. The length of the lifted small tangent vector is given explicitly by

$$|d\vec{p}(s)|_{g(p(s),s)} = [|\vec{p}(s)|ds]_{g(p(s),s)}$$

$$= \left[\left(\delta_{j,k} \frac{dp^j(s)}{ds} \frac{dp^k(s)}{ds} \right)^{1/2} ds \right]_{g(p(s),s)}. \tag{6.67}$$

This equation also describes the result of the lift of the length, $|d\vec{p}(s)|$, of the vector in the base space and time.

Figure 6.11 illustrates this setup for a few points on the path. The base space and time are shown with time in the vertical axis and space in the horizontal axis. Also shown is a path p and the local spaces associated with a few path points.

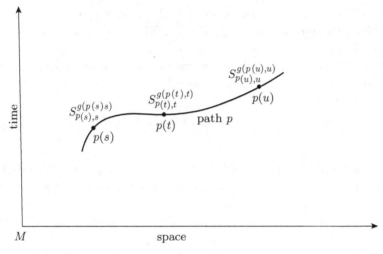

Figure 6.11. Schematic representation of local spaces associated with points of a path in a background space and time. Both the background spaces and the local spaces are Euclidean.

Parallel transport of the differential length $|d\vec{p}(s)|_{g(p(s),s)}$ at $p(s), s$ to a location x, t is expressed by

$$|d\vec{q}(s)|_{g(x,t)} - C_g(x, t; p(s), s)|d\vec{p}(s)|_{g((s),s)}$$

$$= [e^{-\alpha(x,t)+\alpha(y,s)}|d\vec{p}(s)|]_{g(x,t)} \qquad (6.68)$$

$$= [e^{-\alpha(x,t)+\alpha(y,s)}|d\vec{p}(s)|]_{g(x,t)}.$$

The path $q_{g(x,t)}$ is represented by these tangent vectors for all s. The length of the path is represented by the integral

$$\left[\int |d\vec{q}(s)|\right]_{g(x,t)} = \left[e^{-\alpha(x,t)} \int e^{\alpha(p(s),s)}|d\vec{p}(s)|\right]_{g(x,t)}. \qquad (6.69)$$

This description can be extended to differential vectors at any location in $M \times T$, not just at path locations. Let z, s be an arbitrary space and time location in $M \times T$ and $d\vec{z}$ a differential vector at z, s. The length of $d\vec{z}$ is $|d\vec{z}|$.

Lift of the length of the local space at z, s followed by parallel transport to the local reference space at x, t gives the length $[e^{\alpha(z,s)-\alpha(x,t)}|d\vec{z}|]_{g(x,t)}$. This is a numerical quantity in $\bar{R}_{x,t}^{g(x,t)}$. This is the length of the vector $[e^{\alpha(z,s)-\alpha(x,t)}d\vec{z}]_{g(x,t)}$ at the point $z_{g(x,t)}$ in the local space $\mathbb{S}_{x,t}^{g(x,t)}$.

6.4.1 *Relative space sizes*

The relative sizes of the different local spaces can be determined from the metric tensors associated with each of the spaces. These are obtained from the expression for the lifted, transported differential length, $|d\vec{z}|$, as

$$[e^{\alpha(z,s)-\alpha(x,t)}|d\vec{z}|]_{g(x,t)} = [(e^{-2\alpha(x,t)+2\alpha(z,s)}\delta_{j,k}\partial_j z\partial_k z)^{1/2}]_{g(x,t)}.$$
$$(6.70)$$

The metric tensor, $[e^{-2\alpha(x,t)+2\alpha(z,s)}\delta_{j,k}]_{g(x,t)}$, is an expansion of that in Eq. (6.61) as it depends on the times, s and t, and the space locations, z and x. If the exponent is positive, and the local space $\mathbb{S}_{z,s}^{g(z,s)}$ is an expansion of the local space at x,t. If the exponent is negative, the local space at z,s is a contraction of the space at x,t. If the exponent is 0, the two spaces are the same size.

Figure 6.12 is an illustration of the expansion or contraction of the local spaces along a path p as the time increases from s to t, where

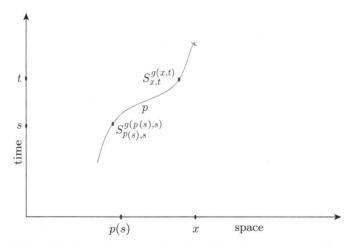

Figure 6.12. The expansion or contraction of the local space at $p(s),s$ relative to the local space at x,t. The size of the $\mathcal{S}$ symbols at $p(s),s$ relative to the size of $\mathcal{S}$ at x,t denotes relative space expansion or contraction. The two symbols at $p(s),s$ denote expansion and contraction relative to that at $p(t) = x$.

$p(t) = x$. The font sizes of $\mathbb{S}$ of the local spaces at $p(s), s$ relative to that at the path point at $p(t), t$ indicate expansion or contraction.

The size relations are an expansion of those described in the previous section in that they depend on time as well as on space. The extent and shape of the regions of expanded and contracted spaces shown in Figure 6.9 depend on time as well as on space locations. The rate of shape change is determined by the gradient potential of α.

So far, no limit is placed on the location of the reference site. It can be anywhere in $M \times T$. However, observers play an important role here. The local mathematics at a reference location is the mathematics that an observer, collocated with the mathematics, uses to construct meaningful theoretical models of the physics and geometry of systems in the base space and time, $M \times T$. As an observer moves along a path in $M \times T$, the mathematics and local space an observer uses for theory descriptions move along the path with the observer.

As an example of this, if the exponent $-\alpha(x, t) + \alpha(p(s), s)$ is positive, an observer at x, t would conclude that the local space at $p(s), s$ is an expansion of the local space at the observers location. If the exponent is negative, the observer would know that the local space at $p(s), s$ is a contraction of the space at x, t.

The reference locations are restricted by the requirement that they are limited to those locations occupiable by an observer. As will be seen later, these locations are restricted to be a small fraction of the cosmological universe.

The special case in which α is independent of space location and depends on time only is of interest. The local spaces at different space locations at the same time are identical. The independence is expressed by $\alpha(y, s) = \alpha(s)$ for all y and $\alpha(x, t) = \alpha(t)$.

The identity follows from the observation that the parallel transport map between two local spaces at different space locations and the same time is the identity map. The local spaces at each space point at the same time can be collapsed into one space for all space locations at a fixed time. One has $\mathbb{S}_{y,s}^{g(y,s)} \to \mathbb{S}_{s}^{g(s)}$ for all y.

The view for an observer at t is determined by the time variation of $\alpha(s)$. If $-\alpha(t) + \alpha(s) < 0$ and $\alpha(s) \to \alpha(t)$ as s increases to t, the observer at t would conclude that the local space, $\mathbb{S}_{s}^{g(s)}$, is smaller than the local space at t and is expanding. If $-\alpha(t) + \alpha(s) > 0$ and $\alpha(s) \to \alpha(t)$ as s increases to t, the observer at t would conclude that the local space, $\mathbb{S}_s^{g(s)}$, is larger than the local space at t and is contracting. At $s = t$, the two spaces are identical.

The rate of expansion or contraction as the time s increases is given by the time derivative of the metric tensors for the local spaces. It is given by

$$\frac{d}{ds} e^{-\alpha(t)+\alpha(s)} \delta_{j,k} = A(s) e^{-\alpha(t)+\alpha(s)} \delta_{j,k}. \qquad (6.71)$$

If $A(s)$ is positive, the spaces are expanding with a velocity of expansion given by the above equation. If $A(s)$ is negative, the spaces are contracting with a negative velocity, also given by the above equation.

There are two views one can take of the local spaces. One view emphasizes the properties of physical systems in a space at a reference location in M. These are theoretical descriptions of the systems and their dynamics and interactions in M. The separation between the systems and their description is emphasized by the lift and parallel transport of system properties and dynamics to the reference space. This view emphasizes theory descriptions of the properties of physical systems and their dynamics as systems in the base space M.[7]

The theoretical description of the motion of a particle in M is an example of this view. Let p be the path of a particle in M. The particle position and velocity at time s are denoted by $p(s)$ and $\vec{p}(s)$ in M. Lift of these quantities to the space at position $p(s)$ gives the position $p(s)_{g(s)}$ and velocity $\vec{p}(s)_{g(s)}$ in the space $\mathbb{S}_{g(s)}^{g(p(s))}$ at time s.

Comparison of these velocities requires parallel transport to a reference location t where the velocities can be compared. The result gives the velocities

$$[\vec{w}(s)]_{g(t)} = [e^{\alpha(s)-\alpha(t)} \vec{p}(s)]_{g(t)}. \qquad (6.72)$$

The velocity, $\vec{w}(s)_{g(t)}$, is the theory description of the motion of a particle on a path p in M. As it is a general description for any path, the description is independent of the relation between s and t. If $s < t$, it is a theory description for the velocity of a particle on the

---

[7]Recall that all these takes place in the arena of local mathematics and effects of the "no information at a distance" principle.

path p in M in the past. If $s > t$, $\vec{w}(s)_{g(t)}$ is a theory description of the future velocity of the particle velocity on the path p in M.

The exponential gives the effect on the velocities of the particle as it moves through spaces of different sizes. If $\alpha(s)$ increases with increasing time, the velocity of the particle increases with time. If $\alpha(s)$ decreases with increasing time, the velocity of the particle decreases as time increases. These velocities represent the particle motion as seen by an observer at t.

The acceleration or deceleration of the particle motion seen by an observer at t is given by[8]

$$\frac{d\vec{w}(s)}{ds} = \frac{d}{ds}(e^{\alpha(s)-\alpha(t)}\vec{p}(s))$$
$$= e^{\alpha(s)-\alpha(t)}\left(\frac{d}{ds} + A(s)\right)\vec{p}(s). \tag{6.73}$$

The component $A(s)\vec{p}(s)$ of the acceleration or deceleration of the particle motion is present even if $\vec{p}(s)$ is constant. It is independent of the mass of the particle. It expresses the acceleration or deceleration of the rate of expansion or contraction felt by the particle as it passes through the different sized local spaces on its path. To save on notation the subscript, $g(x, t)$ has been suppressed.

The component, d/ds, of the acceleration, represents the usual effect of forces acting on the particle motion. If the particle is free with no external forces present, this component of the acceleration or deceleration is zero. The component due to the meaning field derivative, $A(s)$, is still present.

There is no limit on the size of the particles and their motion. The particles can be as small as nucleons. They can be as large or larger than galaxies in the universe. The theory description of these systems and their motion in the space and time geometry is the same.

This view emphasizes the properties of systems within local spaces at different space and time locations in M. The spaces can be cosmological in that they can include all the physical systems in the universe.

---

[8]Since α depends only on time, these results are the same for all space locations. They depend only on time.

The other view describes the effect of the meaning field on the geometries of the spaces at different space and time locations. Each space at time s is a local representation of the universe at cosmological time, s. For an observer at time t, the local space at t is the reference space. The local mathematics and the reference space at t are used by an observer to give a theoretical model of the time-dependent expansion or contraction of the local spaces as models of the expansion or contraction of the universe.

As might be expected, the geometries of the spaces at time s and the reference space at t are different. From the view point of an observer at t, the metric for a local space at s is Euclidean. To see this, let z be any point in $\mathbb{S}_s^{g(s)}$ and $d\vec{z}$ a differential vector at the point z. Since z can be any point in the local space, including the path point, $p(s)$, the scaled length squared of $d\vec{z}$ at location z in the space at s is

$$e^{-2\alpha(t)+2\alpha(s)}|d\vec{z}|^2 = e^{-2\alpha(t)+2\alpha(s)}\delta_{j,k}\partial_j z \partial_k z.$$

The Euclidean nature of the spaces for all times s different from t follows from the observation that the exponent of the scaled length is the same for all locations z in the spaces at times s. This is not the case for the geometry of the space at t. There the exponent of the scaled length depends on the location of the vector in the space. This is shown by the scaled length squared of the vector, $d\vec{p}(s)$, at location, $p(s)$, in the space at t:

$$e^{-2\alpha(t)+2\alpha(s)}|d\vec{p}(s)|^2 = e^{-2\alpha(t)+2\alpha(s)}\delta_{j,k}\partial_j p(s) \partial_k p(s).$$

The time s at which the path point is at location $p(s)$ is the same as the time appearing in the exponent.

The location dependence of the scaled length squared can be made explicit by replacing the time s with another time, u. The scaled path length square, $e^{-2\alpha(t)+2\alpha(u)}\delta_{j,k}\partial_j p(u)\partial_k p(u)$, applies only to the point $p(u)$ in the reference space at t. It is different from the value at location $p(s)$ in the space at t. The essential point is that the same times, s or u, appear both in the location and in the meaning field as in $\alpha(s)$ or $\alpha(u)$.

The description of the differences in the geometries of the spaces applies if the local spaces represent the universe at different times and the times s and t are cosmological times with $s = t = 0$ the time of the big bang. In this case,

$$e^{-2\alpha(t)+2\alpha(s)} \left(\sum_j (dz^j)^2 \right) \tag{6.74}$$

is the metric for the universe at time s relative to that for the universe at time t. If the universe is homogenous and isotropic, as is the case if it is empty or filled with uniform particle dust, the Euclidean coordinates can be replaced by reduced circumpolar coordinates. The metric becomes

$$e^{-2\alpha(t)+2\alpha(s)} r^2 ((d\theta)^2 + (\sin(\theta)d\phi)^2). \tag{6.75}$$

As cosmological time increases, both s and t increase by the same amount. If Δ is a time increase, the size of a space at time $s + \Delta$, relative to the size of the space at $t + \Delta$, is determined by the time scaling factor, $\exp(-2\alpha(t+\Delta)+2\alpha(s+\Delta))(\sum_j (dz^j)^2)$, in the metric.

The time rate of change of the relative size change of the spaces at times s is given by

$$(\partial_s + \partial_t)e^{-2\alpha(t)+2\alpha(s)} \delta_{j,k} = 2(A(s) + A(t))e^{-2\alpha(t)2\alpha(s)} \delta_{j,k}. \tag{6.76}$$

The same result is obtained for the metric as

$$(\partial_s + \partial_t)e^{-2\alpha(t)+2\alpha(s)} \left(\sum_j (dz^j)^2 \right)$$

$$= 2(A(s) + A(t))e^{-2\alpha(t)2\alpha(s)} \left(\sum_j (dz^j)^2 \right). \tag{6.77}$$

These equations show that the reference time, t, of the observer and the local mathematics and space at t are not fixed at a time different from the cosmological time. Mathematics and spaces at times t and s are comoving.

6.5 Space–time

The description of the effects of local mathematics and the presence of a value field in the space–time of special relativity requires several changes from that used for Euclidean space. The underlying manifold M is the four-dimensional Minkowski space or space–time. The metric tensor for the space is $\eta_{\mu,\nu} = \eta_{\mu,\mu} = 1, -1, -1, -1$. Components of points, y, in the space can be represented as $y = \{y^\nu | \nu = 0, 1, 2, 3\} = ct$, $\mathbf{y}$, where $y^0 = ct$. Here c is the speed of light and t is the time.

Paths in M are either timelike, lightlike, or spacelike. Timelike paths are those for which the length elements are all positive. Length elements for lightlike paths are 0 and negative for spacelike paths. In the first part of this section, paths will be treated as timelike. The discussion will then be extended to lightlike paths.

As is well known, timelike paths describe the motion of physical systems with positive rest mass. Here paths will be treated as geometric objects independent of the mass of any systems moving along the path. The only field present is the meaning or value field, α.

The fiber bundle is similar to that for Euclidean space. It is given in Eq. (6.1). The fiber F is defined by

$$F = \cup_r \cup_S S^r \cup \mathbb{S}^r. \tag{6.78}$$

As before, the union over S is over all mathematical systems that include numbers in their description. $\mathbb{S}^r$ is a scaled four-dimensional space–time. For each r, a vector, $\vec{v}$, in M has an image vector, $\vec{v}_r$, in $\mathbb{S}^r$. The metric tensor for $\mathbb{S}^r$ is

$$[\eta_{\nu,\nu}]_r = 1_r, -1_r, -1_r, -1_r. \tag{6.79}$$

Here 1_r and -1_r are numbers in $\bar{R}^r$ with number values, 1 and -1.

As was the case for Euclidean space, the components of the fiber at y are given by the inverse of the projection operator. These are shown by

$$\pi_g^{-1}(y) = F_y^{g(y)} = \cup_S S_y^{g(y)} \cup \mathbb{S}_y^{g(y)}. \tag{6.80}$$

Since y is a location in space–time, the fiber, $F_y^{g(y)}$, and its components have a location in time as well as in space.

The g field is defined from a real scalar field, α defined on M where for each point, z, on M

$$g(z) = e^{\alpha(z)}. \tag{6.81}$$

The fields α and g are similar to the definitions for time and Euclidean space in that these fields depend on space and time.

The fibers at the different locations in space–time of Eq. (6.80) show that each point, $x = t, \mathbf{x}$, of space–time is associated with a local space–time, $\mathbb{S}_x^{g(x)}$. As was the case for Euclidean space and time, there are separate space–times for points in M that differ only in time. The image space–time associated with another point $y = t', \mathbf{x}$ at the same space location as x but at a different time is $\mathbb{S}_y^{g(y)}$. This image space–time is different from the image space–time at x. The difference is shown by the differences in the value of α at y and x.

Figure 6.13 illustrates this setup in space–time. Two of an infinite number of image space–times at points in M with the same space location but different times are shown.

As was done in Euclidean space, lift of whole paths and their tangent vectors to a local space–time at an arbitrary location y in M and their parallel transports from one location to another are described first. This description is followed by a representation of path lifts in which the bundle of path tangent vectors is treated as a section on the fiber bundle. In this case, the lift location for each tangent vector is restricted to be the local space–time at the same location as the tangent vector in M. Since the main interest is in the

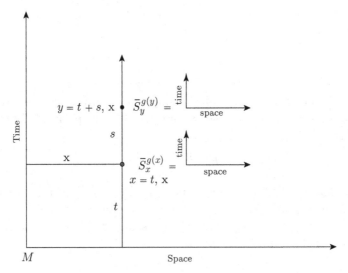

Figure 6.13. Illustration of two of the many image space–times at the same space location, $\mathbf{x}$, but different times, t and $s + t$, in the space–time, M. The image space–time diagrams are also shown.

lift of tangent vectors to space–times collocated with the vectors, the discussion of lift to a point y will be brief.

The image path tangent vectors are defined by the action of chart maps on the corresponding objects on the path in M. For each path point, $p(s)$, the image of the tangent vector $d\vec{p}(s)$ is $[d\vec{p}(s)]_{g(p(s))}$ in $\mathbb{S}_{p(s)}^{g(p(s))}$ as in .

$$[d\vec{p}(s)]_{g(p(s))} = \rho_{g(p(s))}(d\vec{p}(s)). \tag{6.82}$$

For each time, s, the location of the image tangent vector in the local space is the image point, $p(s)_{g(p(s))}$.

For any location, y in M the local velocity of light, as a number in $\bar{R}_y^{g(y)}$, is the number, $c_{g(y)}$. This number has value c, about 300,000 km/s.

6.5.1   *Transport of local images of whole paths*

Let p be a path in the base space–time M Lift of this whole path to a local space–time at y gives the image path, $p_{g(y)}$, in the space–time $\mathbb{S}_y^{g(y)}$. Parallel transportation of the image path, $p_{g(y)}$, to a path, $q_{g(x)}$, in $\mathbb{S}_x^{g(x)}$ at a location, x, on M is implemented by the connection, $C_g(x, y)$, where

$$q_{g(x)} = C_g(x, y)p_{g(y)} = [e^{-\alpha(x)+\alpha(y)}p]_{g(x)} = [e^{-\alpha(x)+\alpha(y)}]_{g(x)}p_{g(x)}. \tag{6.83}$$

Since parallel transports are path preserving and path value changing, $q_{g(x)}$ is the same path in $\mathbb{S}_x^{g(x)}$ as $p_{g(y)}$ is in $\mathbb{S}_y^{g(y)}$, but its path value has changed.

Here $p_{g(x)}$ is the image in $\mathbb{S}_x^{g(x)}$ of p in M. Both $p_{g(y)}$ and $p_{g(x)}$ have the same path values, p, but, as Eq. (6.83) shows, they are different paths. This difference can also be expressed by the chart relations:

$$p_{g(x)} = \rho_{g(x)}\rho_{g(y)}^{-1}(p_{g(y)}). \tag{6.84}$$

This equation shows that charts preserve image path values but not image paths. The parallel transport connection, $C_g(x, y)$, of Eq. (6.83) preserves image paths but not the values.

Parallel transport of path images as mathematical representations of paths in different local space–times are mathematical transports of path images from a local space–time at one location to another local space–time at another location. The position of the path in the space–time, M, is unchanged. Parallel transport changes the location of a mathematical representation of a path. It is not related to the physical translation of a path in M.

Let $p_{g(x)}$ be the image of the path in the local space–time at x. The relations between the path images, $p_{g(y)}$, $p_{g(x)}$, and $q_{g(x)}$, depend on the relative values of $g(y)$ and $g(x)$. The path, $q_{g(x)} = [(g(y)/g(x))p]_{g(x)}$, is the same path in the local space–time at x as $p_{g(y)}$ is at y. If the value factor, $g(x)$, is greater than 1, then the path, $p_{g(x)}$, is an expansion of the path, p, in M. If $g(x)$ is less than 1, then $p_{g(x)}$ is a contraction of p. Note that parallel transportation preserves the timelike or lightlike nature of the path being transported.

Figure 6.14 illustrates the differences between these paths. Space–time, M, is shown as a two-dimensional z-t plane. The third dimension, r, denotes value factors. Two images of M are shown, $\mathbb{S}_x^{g(x)}$, with a value factor $g(x) > 1$ and $\mathbb{S}_y^{g(y)}$ with $g(y) < 1$. Dotted lines show that $\mathbb{S}_y^{g(y)}$ is below the M plane at $r = 1$. Chart images, $p_{g(y)}$ and $p_{g(x)}$, are shown as paths in the images of M. The path $q_{g(x)}$ is shown along with its relation to $p_{g(y)}$ and $p_{g(x)}$.

The length, $L(p_{g(y)})$, of the lifted path $p_{g(y)}$ in $\mathbb{S}_y^{g(y)}$ is a numerical quantity in $\bar{R}_y^{g(y)}$. This number is the same as $[L(p)]_{g(y)}$. This numerical quantity is also the lift to $\bar{R}_y^{g(y)}$ of the length of p in the base space–time.

Parallel transport of the length, $L(p_{g(y)}) = [L(p)]_{g(y)}$ of $p_{g(y)}$, to x is the number, $[e^{-\alpha(x)+\alpha(y)}L(p)]_{g(x)}$. This is the same number as is the length, $L(q_{g(x)})$, of the transported path. This can be seen from

$$L(q_{g(x)}) = [L(e^{-\alpha(x)+\alpha(y)}p)]_{g(x)} = [e^{-\alpha(x)+\alpha(y)}]_{g(x)}]L(p_{g(x)}). \quad (6.85)$$

The length of $q_{g(x)}$ is

$$L(q_{g(x)}) = [L(q)]_{g(x)} = \int_0^t \left[\left(\eta_{\mu,\nu} \frac{dq^\mu(s)}{ds} \frac{dq^\nu(s)}{ds} \right)^{1/2} ds \right]_{g(x)}$$

$$= [L(e^{-\alpha(x)+\alpha(y)}p)]_{g(x)} = [e^{-\alpha(x)+\alpha(y)}L(p)]_{g(x)}. \quad (6.86)$$

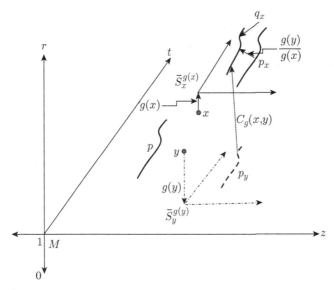

Figure 6.14. Illustration of two images of space–time, M, and paths. Space–time is shown as the z–t plane. The third dimension, r, denotes value factors. The image, $\mathbb{S}_y^{g(y)}$, and path, $p(y)$, are shown with dot-dash lines to indicate that it is below the z-t plane with $g(y) = r < 1$. For the space $\mathbb{S}_x^{g(x)}$, one finds $g(x) > 1$. The paths, p_y and p_x, are chart images of p as $p_x = \rho_{g(x)}(p)$ and $p_y = \rho_{g(y)}(p)$. Paths, not path values, are shown. This is the reason for showing q_x as a contraction of p_x as in $q_x = (g(y)/g(x))p_x$. The path q_x is the same as the path p_y. To avoid the resulting confusion between paths, the path q_x is shown at a separate location from that of p_x. To save on notation, path subscripts are shortened from $g(x)$ and $g(y)$ to x and y.

These results show that the length of the transported path is the same number as is the transported length of a path. As was noted before, this commutativity holds because of the structure of the length integrand. This can be seen explicitly by replacing $dq^\mu(s)/ds$ and $dq^\nu(s)/ds$ with $e^{\alpha(y)-\alpha(x)}(dp^\mu(s)/ds)$ and $e^{\alpha(y)-\alpha(x)}(dp^\nu(s)/ds)$ in the integrand in the above equation.

Note that for $\mu = 0$,

$$\left[\frac{dp^\mu(s)}{ds}\right]_{g(y)} = c_{g(y)} \quad \text{and} \quad \left[\frac{dp^\mu(s)}{ds}\right]_{g(x)} = c_{g(x)}. \tag{6.87}$$

Parallel transport of $c_{g(y)}$ to x gives the number

$$b_{g(x)}(y) = [e^{-\alpha(x)+\alpha(y)}c]_{g(x)}. \tag{6.88}$$

This is the same numerical quantity at x as $c_{g(y)}$ is at y. It is the scaled speed of light resulting from the parallel transport of $c_{g(y)}$ to x. The presence of the exponential factor shows that the numbers have different values.

The fact that parallel transport of the light velocity gives a number with a value different from c does not conflict with the fact that the velocity of light is a constant in all inertial frames in special relativity. At each point, y, the description of special relativity is contained in the local mathematics and space–time at y. The number $c_{g(y)}$ as the velocity of light is the same in all inertial frames at y.

However, as seen above, $c_{g(y)}$ does not represent the velocity of light in the local mathematics and space–time at x. The number $b_{g(x)}(y)$ represents the velocity of light in the local mathematics and space–time at y as seen by an observer using the local mathematics and space–time at x. All equations in special relativity describing the properties of systems at y are invariant under parallel transport to x. This is the case even though $b_{g(x)}(y)$ represents the light velocity in the transported equations.

6.5.2   *Localized paths as tangent vector sections*

So far, the lifting of complete paths to local images of space–time has been considered. However, paths are extended objects. Lifting an extended object to a local space–time is not consistent with the viewpoint taken from gauge theory. Instead, lifting of local objects in M to local images of space–time is to be preferred.

Representation of the path p as a bundle of small tangent vectors, $d\vec{p}(s)$, in M is followed by a lift of each tangent vector, $d\vec{p}(s)$, to a vector, $[d\vec{p}(s)]_{g(p(s))}$, in the local space–time, $\mathbb{S}_{p(s)}^{g(p(s))}$. Parallel transport of the lifted tangent vectors to a reference space–time image at a single location is needed to determine various path properties. Path length is a simple example. As has been noted before, the integral used to define path lengths requires that the integrands all belong to a real number structure at a single location.

Let x be an arbitrary location in M. Parallel transport of the lifted tangent vector, $d\vec{p}(s)_{g(p(s))}$, in $\mathbb{S}_{p(s)}^{g(p(s))}$ to x gives the vector $d\vec{q}(s)_{g(x)}$ in the reference space–time $\mathbb{S}_x^{g(x)}$. The transported vector is defined by

$$d\vec{q}(s)_{g(x)} = C_g(x, p(s))d\vec{p}(s)_{g(p(s))} = [e^{-\alpha(x)+\alpha(p(s))}d\vec{p}(s)]_{g(x)}.$$
$$(6.89)$$

The same relation holds for each of the four components of the vectors. One has

$$[dq^\mu(s)]_{g(x)} = [e^{-\alpha(x)+\alpha(p(s))}dp^\mu(s)]_{g(x)}.$$
$$(6.90)$$

Figure 6.15 shows the relations between tangent vectors on the path p in M, the lifted tangent vectors in two local space–time images, and the vectors resulting from parallel transport of the lifted

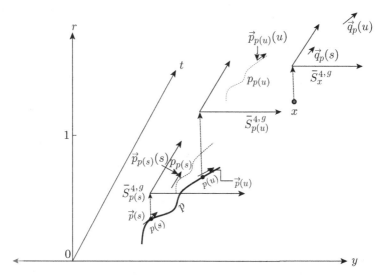

Figure 6.15. Representation of the parallel transport of just one tangent vector in $\mathbb{S}_{p(s)}^{g(p(s))}$ and one in $\mathbb{S}_{p(u)}^{g(p(u))}$ to q_p tangent vectors in $\mathbb{S}_x^{g(x)}$. To save clutter in the figure, the arguments of the g superscript have been suppressed. Space–time is indicated by the y-t plane. The third dimension, r, is a value axis. The heights of the image spaces above the y-t plane, shown by the dashed arrows, indicate the relative magnitude of the value factors, $g(p(s)) < g(x)$ and $g(p(u)) > g(x)$. The lengths of the vectors, $\vec{q}_p(s)$ and $\vec{q}_p(u)$, compared to the lengths of $\vec{p}_{g(p(s))}(s) = \vec{p}(s)_{g(p(s))}$ and $\vec{p}_{g(p(u))}(u) = \vec{p}(u)_{g(p(u))}$, show the changes in vector values. This is why the tangent vector value, $\vec{q}_p(s)$, is shorter than $\vec{p}_{p(s)}(s)$ and $\vec{q}_p(u)$ is longer than $\vec{p}_{p(u)}(u)$. To save space in the figure, the subscripts, $p, p(s)$ and $p(u)$, on paths and vectors denote $g(x), g(p(s))$, and $g(p(u))$. The figure shows the path, p, and its images as timelike world lines. Also the point, x, is timelike relative to all points in p.

vectors to a common location, x. As a visualization aid, images of the lifted whole paths are shown with light dashed lines.

The length element, $|d\vec{q}(s)|_{g(x)}$, of the path, $q_{g(x)}$, is given by

$$|d\vec{q}(s)|_{g(x)} = \left[\left(\eta_{\mu,\nu}\frac{dq^\mu(s)}{ds}\frac{dq^\nu(s)}{ds}\right)^{1/2}ds\right]_{g(x)}$$

$$= \left[e^{-\alpha(x)+\alpha(p(s))}\left(\eta_{\mu,\nu}\frac{dp^\mu(s)}{ds}\frac{dp^\nu(s)}{ds}\right)^{1/2}ds\right]_{g(x)}$$

$$= [e^{-\alpha(x)+\alpha(p(s))}|d\vec{p}(s)|]_{g(x)}. \tag{6.91}$$

The subscript $g(x)$ emphasizes the membership of the vectors and numbers in $\mathbb{S}_x^{g(x)}$ and $\bar{R}_x^{g(x)}$.

The path length from time 0 to time t is given by the integral

$$[c\Delta\tau_0^t]_{g(x)} = [L(q)_0^t]_{g(x)} = \left[\int_0^t\left(\eta_{\mu,\nu}\frac{dq^\mu(s)}{ds}\frac{dq^\nu(s)}{ds}\right)^{1/2}ds\right]_{g(x)}$$

$$= \left[e^{-\alpha(x)}\int_0^t e^{\alpha(p(s))}\left(\eta_{\nu,\mu}\frac{dp^\nu(s)}{ds}\frac{dp^\mu(s)}{ds}\right)^{1/2}ds\right]_{g(x)}$$

$$= \left[e^{-\alpha(x)}\int_0^t e^{\alpha(p(s))}|d\vec{p}(s)|\right]_{g(x)}. \tag{6.92}$$

The equation shows that $[L(q)_0^t]_{g(x)}$ also represents the proper time interval along the path, $q_{g(x)}$, as measured by a clock moving along the path. This equation and Eq. (6.91) show that the proper time along the path depends on the variations in α. As will be seen from the geodesic equation for $q_{g(x)}$, the four-component gradient field, $\vec{A}(p(s))$, of $\alpha(p(s))$ is the main driver of the rate of change in $|d\vec{q}(s)|_{g(x)}$.

6.5.2.1   *Geodesics for localized paths*

Geodesic equations give the description of the path for which the length of the path between two points is an extremum. In space–time, the extremum is a maximum. It is the path for which the proper time elapsed for a clock moving along the path is a maximum [72].

The path $q_{g(x)}$ in $\mathbb{S}_x^{g(x)}$ is represented by the tangent vector bundle,

$$\{d\vec{q}(s)_{g(x)} : 0 < s < t\} = \{[e^{-\alpha(x)+\alpha(p(s))}d\vec{p}(s)]_{g(x)} : 0 < s < t\},$$

in $\mathbb{S}_x^{g(x)}$. The extremal path length value for this bundle is represented by the extremum of Eq. (6.92). This is represented by

$$[ext_p L(q)]_{g(x)} = \left[ext_p e^{-\alpha(x)} \int_0^t e^{\alpha(p(s))} |d\vec{p}(s)| \right]_{g(x)}$$

$$= \left[e^{-\alpha(x)} ext_p \int_0^t e^{\alpha(p(s))} |\vec{p}(s)| ds \right]_{g(x)}$$

$$= \left[e^{-\alpha(x)} ext_p \int_0^t e^{\alpha(p(s))} \left(\eta_{\nu,\nu} \left(\frac{dp^{\nu}(s)}{ds} \right)^2 \right)^{1/2} ds \right]_{g(x)}.$$

$$\tag{6.93}$$

Note that the extremum is obtained for path length values.[9]

The extremum is obtained by application of the Euler–Lagrange equations to the integrand, I, of Eq. (6.93). These equations require that, for each ν, I must satisfy the equation

$$\frac{dI(s)}{dp^{\nu}(s)} - \frac{d}{ds}\frac{dI(s)}{dp^{\nu}(s)/ds} = 0. \tag{6.94}$$

The first term is

$$\frac{dI(s)}{dp^{\nu}(s)} = e^{\alpha(p(s))} A_{\nu}(p(s)) |\vec{p}(s)|. \tag{6.95}$$

Here

$$A_{\nu}(p(s)) = \frac{d\alpha(p(s))}{dp^{\nu}(s)} \tag{6.96}$$

---

[9]Summation convention is sometimes abused to save space in formulas. For example, the integrand of Eq. (6.93) should involve $\eta_{\mu,\nu}(dp^{\mu}(s)/ds)(dp^{\nu}(s)/ds)$ instead of $\eta_{\nu,\nu}(dp^{\nu}(s)/ds)^2$. This should be clear from the context.

and

$$|\vec{p}(s)| = \left[\eta_{\mu,\mu}\frac{dp^{\mu}(s)}{ds}\frac{dp^{\mu}(s)}{ds}\right]^{1/2}. \tag{6.97}$$

For the second term,

$$\frac{dI(s)}{d(dp^{\nu}(s)/ds)} = e^{\alpha(p(s))}\eta_{\nu,\nu}\frac{dp^{\nu}(s)}{ds}\frac{1}{|\vec{p}(s)|} \tag{6.98}$$

with no sum over ν.

Carrying out the d/ds derivative gives the result that

$$\frac{d}{ds}\frac{dI(s)}{d(dp^{\nu}(s)/ds)} = e^{\alpha(p(s))}\left[A_{\mu}(p(s))\frac{dp^{\mu}(s)}{ds}\eta_{\nu,\nu}\frac{dp^{\nu}(s)}{ds}\frac{1}{|\vec{p}(s)|}\right.$$
$$\left. + \eta_{\nu,\nu}\frac{d}{ds}\left(\frac{dp^{\nu}(s)}{ds}\frac{1}{|\vec{p}(s)|}\right)\right]. \tag{6.99}$$

Here use has been made of the chain rule for differentiation as in

$$\frac{d\alpha(p(s))}{ds} = \frac{d\alpha(p(s))}{dp^{\mu}(s)}\frac{dp^{\mu}(s)}{ds}.$$

Also sum over μ is implied.

Combining Eqs. (6.95) and (6.99) in Eq. (6.94) and rearranging terms give

$$\eta_{\nu,\nu}\frac{d}{ds}\left(\frac{dp^{\nu}(s)}{ds}\frac{1}{|\vec{p}(s)|}\right)$$
$$= A_{\nu}(p(s))|\vec{p}(s)| - A_{\mu}(p(s))\frac{dp^{\mu}(s)}{ds}\eta_{\nu,\nu}\frac{dp^{\nu}(s)}{ds}\frac{1}{|\vec{p}(s)|}. \tag{6.100}$$

The exponential factor, common to both sides of the equation, has been canceled. Dividing both sides of the equation by $|\vec{p}(s)|$ gives

$$\eta_{\nu,\nu}\frac{1}{|\vec{p}(s)|}\frac{d}{ds}\left(\frac{dp^{\nu}(s)}{ds}\frac{1}{|\vec{p}(s)|}\right)$$
$$= A_{\nu}(p(s)) - A_{\mu}(p(s))\frac{dp^{\mu}(s)}{ds}\eta_{\nu,\nu}\frac{dp^{\nu}(s)}{ds}\frac{1}{|\vec{p}(s)|^{2}}. \tag{6.101}$$

The form of this equation shows that it can be simplified by change of the variable s as time to τ as proper time. Use of the relation

$$cd\tau = |\vec{p}(s)|ds \tag{6.102}$$

gives

$$\eta_{\nu,\nu}\frac{d^2}{c^2d\tau^2}dp^\nu(\tau) = A_\nu(p(\tau)) - \frac{A_\mu(p(\tau))}{c^2}\frac{dp^\mu(\tau)}{d\tau}\eta_{\nu,\nu}\frac{dp^\nu(\tau)}{d\tau}. \tag{6.103}$$

The metric tensor can be removed from the left-hand term of the equation by multiplying both sides by $\eta^{\nu,\nu}$ and c^2 to get

$$\frac{d^2}{d\tau^2}p^\nu(\tau) = \eta^{\nu,\nu}c^2 A_\nu(p(\tau)) - A_\mu(p(\tau))\frac{dp^\mu(\tau)}{d\tau}\frac{dp^\nu(\tau)}{d\tau}. \tag{6.104}$$

The fact that $\eta^{\mu,\nu}\eta_{\nu,\alpha} = \delta_\alpha^\mu = \text{diag}(1,1,1,1)$ has been used. Sum over μ is implied.

There are four of these equations, one for each value of ν. The equations are also coupled to one another. This is seen by the presence of $dp^\mu(\tau)/d\tau$ for all μ in each of the equations.

The geodesic equations for the numbers and vectors are obtained by appending the subscript, $g(x)$, to the factors in Eq. (6.104). These equations should be the same because the derivations from the Euler–Lagrange equations take place in mathematical structures at one location[10] x.

The equations have the good property of being independent of the location x of the local space–time containing the path $q_{g(x)}$ with path value, q. If y is another location, the same equations in Eq. (6.104) give the conditions that $p_{g(y)}$ must satisfy so that the length value, of the path $q_{g(y)}$ in $\mathbb{S}_y^{g(y)}$, is an extremum.

The geodesic equations for the path p give the conditions that the path, p, in M must satisfy to maximize the length of the path, q, in a local space–time. One can use the dependence of the components of the path, q, on the components of p to obtain equations for q from the Euler–Lagrange equations for p.

---

[10]The distinction between mathematical entities, such as numbers and vectors, and their values can be ignored as long as one restricts the description to the local mathematics at one location.

One begins by use of Eq. (6.89) and the relation between each component of $\vec{q}(s)$ and $\vec{p}(s)$. This is given by

$$\frac{dq^\nu(s)}{ds} = e^{-\alpha(x)+\alpha(p(s))}\frac{dp^\nu(s)}{ds}. \tag{6.105}$$

This equation relates the vector value components, $dq^\nu(s)/ds$, to the components $dp^\nu(s)/ds$ in the local space–time at x. The vector equation obtained by adding the subscript $g(x)$ to both sides is also valid.

Use of this equation and

$$|\vec{q}(s)| = e^{-\alpha(x)+\alpha(p(s))}|\vec{p}(s)| \tag{6.106}$$

gives

$$\frac{dq^\nu(s)}{ds}\frac{1}{|\vec{q}(s)|} = \frac{dp^\nu(s)}{ds}\frac{1}{|\vec{p}(s)|}. \tag{6.107}$$

The terms of Eq. (6.101) containing $A_\nu(p(s))$ and $A_\mu(p(s))$ need to be converted to terms depending on $q(s)$. This can be done by use of the relation

$$\frac{d}{ds}\alpha(p(s)) = \frac{d}{dp^\mu(s)}\alpha(p(s))\frac{dp^\mu(s)}{ds} = A_\mu(p(s))\frac{dp^\mu(s)}{ds}$$

$$= \frac{d}{dq^\mu(s)}\alpha(p(s))\frac{dq^\mu(s)}{ds}. \tag{6.108}$$

Use of Eq. (6.105) and rearrangement of terms give

$$\sum_\mu \left\{ A_\mu(p(s)) - e^{-\alpha(x)+\alpha(p(s))}\frac{d}{dq^\mu(s)}\alpha(p(s)) \right\}\frac{dp^\mu(s)}{ds} = 0. \tag{6.109}$$

One can use the observations that, for each μ, the bracketed factors are independent of one another and $dp^\mu(s)/ds \neq 0$ to obtain

$$A_\mu(p(s)) = e^{-\alpha(x)+\alpha(p(s))}\frac{d}{dq^\mu(s)}\alpha(p(s)). \tag{6.110}$$

Define $A_\mu(q(s))$ by

$$A_\mu(q(s)) = \frac{d}{dq^\mu(s)}\alpha(p(s)). \tag{6.111}$$

The relation between $A_\mu(q(s))$ and $A_\mu(p(s))$ is obtained from Eq. (6.110) as

$$A_\mu(p(s)) = A_\mu(q(s))e^{-\alpha(x)+\alpha(p(s))}. \qquad (6.112)$$

This definition can be used to express the Euler–Lagrange equations in terms of $q(s)$ and its derivatives. One obtains

$$\frac{1}{|\vec{q}(s)|}\frac{d}{ds}\left(\frac{dq^\nu(s)}{ds}\frac{1}{|\vec{q}(s)|}\right)$$
$$= \eta^{\nu,\mu}A_\mu(q(s)) - A_\mu(q(s))\frac{dq^\mu(s)}{ds}\frac{dq^\nu(s)}{ds}\frac{1}{|\vec{q}(s)|^2}. \qquad (6.113)$$

Equations (6.91)–(6.107) have been used to obtain this result.

These equations are the geodesic equations for each of the four vector value components of $\vec{q}(s)$. Sum over μ is implied. One sees that the geodesic equations for the vectors, $\vec{q}(s)_{g(x)}$, and vector values, $\vec{q}(s)$, have the same functional form as the equations for $\vec{p}(s)$ and $\vec{p}(s)_{g(x)}$. This results from the fact that converting the factors in each of the terms in the equations for q to terms in the equations for p cancels out the value factors. For example,

$$A_\nu(q(s))|\vec{q}(s)| = e^{\alpha(x)-\alpha(p(s))}A_\nu(p(s))e^{-\alpha(x)+\alpha(p(s))}|\vec{p}(s)|. \qquad (6.114)$$

The path p is a path in the base space–time M. The properties of this path are defined from the requirement that it is such that the length for the path q in the local space–time at x is an extremum. The similarity between the equations for p and q shows that the path, p, has the same properties as the path value q has in $\mathbb{S}_x^{g(x)}$.

The form of these geodesic equations for the local space–times shows that the geometry of the local space–times differs from that of M. For M, the geodesic equations are those of a straight line. No dependence on $\vec{A}$ is present.

It follows that the local spaces are not chart images of M. Nevertheless, the initial representation of the local spaces as chart images provides a good background for seeing how the local geometries are altered by representing tangent vectors of paths in M as sections on the fiber bundles followed by parallel transport of the lifted tangent vectors to a reference space for assembly into a path. The difference between the geometry of M and the geometries of the local spaces is addressed later on in Section 6.5.3.

The effects of the variations in the α field show up in the presence of the gradient components, $\vec{A}(q(s))$. If α is such that the values of $\alpha(p(s))$ are the same for all s, then $\vec{A}(q(s)) = 0$ and Eqs. (6.104) and (6.113) are geodesics for straight paths. Both M and the local space–times have the same geometry.

For the time component with $\eta^{0,0} = 1$, Eq. (6.113) shows that

$$
\frac{d}{ds}\left(\frac{dq^0(s)}{ds}\frac{1}{|\vec{q}(s)|}\right)
$$
$$
= A_0(q(s))|\vec{q}(s)| - A_\mu(q(s))\frac{dq^\mu(s)}{ds}\frac{dq^0(s)}{ds}\frac{1}{|\vec{q}(s)|}.
$$

(6.115)

For the space components with $\eta^{j,j} = -1$ and $j = 1, 2, 3$, Eq. (6.113) shows that

$$
\frac{d}{ds}\left(\frac{dq^j(s)}{ds}\frac{1}{|\vec{q}(s)|}\right)
$$
$$
= -A_j(q(s))|\vec{q}(s)| - A_\mu(q(s))\frac{dq^\mu(s)}{ds}\frac{dq^j(s)}{ds}\frac{1}{|\vec{q}(s)|}.
$$

(6.116)

The corresponding equations for the path p are obtained from Eq. (6.104). They are

$$
\frac{d}{ds}\left(\frac{c}{|\vec{p}(s)|}\right) = A_0(p(s))|\vec{p}(s)| - A_\mu(p(s))\frac{dp^\mu(s)}{ds}\frac{c}{|\vec{p}(s)|}
$$

(6.117)

for the time component and

$$
\frac{d}{ds}\left(\frac{dp^j(s)}{ds}\frac{1}{|\vec{p}(s)|}\right) = - A_j(p(s))|\vec{p}(s)|
$$
$$
- A_\mu(p(s))\frac{dp^\mu(s)}{ds}\frac{dp^j(s)}{ds}\frac{1}{|\vec{p}(s)|}
$$

(6.118)

for the three $j = 1, 2, 3$ space components. For the time component, $dp^0(s)/ds = c$ was used.

The geodesic equations can be given in an equivalent form that is based on the gamma function of special relativity. This follows from

$$\gamma(p(s)) = \left\{ 1 - \frac{\vec{v}(p(s))^2}{c^2} \right\}^{-1/2} \tag{6.119}$$

and the relation between $|\vec{p}(s)|$ and the gamma function as

$$|\vec{p}(s)| = \left\{ \eta_{\mu,\nu} \frac{d^\mu p}{ds} \frac{d^\nu p}{ds} \right\}^{1/2} = \{c^2 - \vec{v}(p(s))^2\}^{1/2} = \frac{c}{\gamma(p(s))}. \tag{6.120}$$

Use of these results and the fact that $d^0 p(s)/ds = c$ enable one to express the geodesic equations as equations for the gamma function. For the time component, one has

$$\frac{d}{ds}(\gamma(p(s))) = cA_0(p(s)) \left(\frac{1}{\gamma(p(s))} - \gamma(p(s)) \right)$$
$$- \sum_{\mu \neq 0} A_\mu(p(s)) \frac{dp^\mu(s)}{ds}. \tag{6.121}$$

This equation is obtained by separating out the time component of $A_\mu(p(s))$ from the μ sum of Eq. (6.118). For the space components, one obtains

$$\frac{d}{ds} \left(\frac{dp^j(s)}{ds} \frac{\gamma(p(s))}{c} \right) = -A_j(p(s)) \frac{c}{\gamma(p(s))}$$
$$- A_\mu(p(s)) \frac{dp^\mu(s)}{ds} \frac{dp^j(s)}{ds} \frac{\gamma(p(s))}{c}. \tag{6.122}$$

As was the case for the geodesic equation for $\vec{p}$ in Eq. (6.104), these equations are coupled differential equations. This makes them difficult to solve for general vector fields $\vec{A}$ that depend on space and time.

The relations between the gamma function and particle kinetic energy and momentum, as

$$E = (\gamma - 1)mc^2 \tag{6.123}$$

and

$$P = (\gamma^2 - 1)^{1/2}mc, \tag{6.124}$$

suggest that Eqs. (6.121) and (6.122) show the effect of the $\vec{A}$ field on particle motion. This is not correct. For example, the equations for the space–time dependence of the gamma function on $\vec{A}$ are independent of the particle mass.

In essence, the problem is that $\vec{A}$ is not a potential field in a space–time with a fixed geometry. If that were the case, then it would affect particle motion. Instead $\vec{A}$ is a field that determines the geometries of the local space–times. The space–time dependence of the gamma function in the above equations reflects the space–time dependence of the geometries of the local spaces.

6.5.2.2 *Special case: $\vec{A}$ depends only on time*

One can get a partial feeling for the properties of Eqs. (6.121) and (6.122) by considering a special case where $\vec{A}$ depends on time only. In this case, $A_\mu(p(s)) = 0$ for $\mu = 1, 2, 3$ and $A_0(p(s)) \neq 0$.

For the time component, Eq. (6.121) becomes

$$\frac{d}{ds}(\gamma(p(s))) = cA_0(p(s)) \left(\frac{1}{\gamma(p(s))} - \gamma(p(s)) \right)$$
$$= cA_0(p(s)) \frac{(1 - \gamma(p(s))^2)}{\gamma(p(s))}. \tag{6.125}$$

For the space components, Eq. (6.122) gives

$$\frac{d}{ds}\left(\frac{dp^j(s)}{ds} \gamma(p(s)) \right) = -cA_0(p(s)) \frac{dp^j(s)}{ds} \gamma(p(s)). \tag{6.126}$$

Expansion of the left-hand derivative and use of Eq. (6.125) give

$$\frac{d^2 p^j(s)}{ds^2} \gamma(p(s)) + \frac{dp^j(s)}{ds} \frac{d\gamma(p(s))}{ds}$$
$$= \frac{d^2 p^j(s)}{ds^2} \gamma(p(s)) + \frac{dp^j(s)}{ds} cA_0(p(s)) \left(\frac{1}{\gamma(p(s))} - \gamma(p(s)) \right)$$
$$= -cA_0(p(s)) \frac{dp^j(s)}{ds} \gamma(p(s)). \tag{6.127}$$

The last equality is the right-hand term of Eq. (6.126). Removal of identical terms from the right- and left-hand sides of the equation gives

$$\frac{d^2 p^j(s)}{ds^2} \gamma(p(s)) + \frac{dp^j(s)}{ds} cA_0(p(s)) \left(\frac{1}{\gamma(p(s))} \right) = 0 \qquad (6.128)$$

or

$$\frac{d^2 p^j(s)}{ds^2} = -\frac{dp^j(s)}{ds} cA_0(p(s)) \left(\frac{1}{\gamma(p(s))} \right)^2. \qquad (6.129)$$

Division of both sides of Eq. (6.128) by $\gamma(p(s))$ gives this result.

This equation has satisfactory properties. If $A_0(p(s)) = 0$, then $\frac{d^2 p^j(s)}{ds^2} = 0$. If $\gamma(p(s))^2$ is very large, then $\frac{d^2 p^j(s)}{ds^2}$ is very small. In the limit, $\gamma(p(s) = \infty$, $\frac{d^2 p^j(s)}{ds^2} = 0$.

Equation (6.126) can be integrated. Let

$$w^j(s) = \frac{dp^j(s)}{ds} \gamma(p(s)). \qquad (6.130)$$

Use of this substitution in Eq. (6.126) gives

$$\frac{dw^j(s)}{ds} = -cA_0(p(s))w^j(s). \qquad (6.131)$$

For time t later than s, one has

$$w^j(t) = w^j(s) \exp \left(-\int_s^t cA_0(p(u)) \right) du$$

or

$$\frac{dp^j(t)}{dt} \gamma(p(t)) = \frac{dp^j(s)}{ds} \gamma(p(s)) \exp \left(-\int_s^t cA_0(p(u)) du \right). \qquad (6.132)$$

This equation has satisfactory properties. If $\gamma(p(s))$ is large, then some of the component velocities, $dp^j(s)/ds$, must be close to the light velocity, c. If the exponent is positive, then almost all of the increase of $(dp^j(t)/dt)\gamma(p(t))$ over the value of $(dp^j(s)/ds)\gamma(p(s))$ comes from an increase in the value of $\gamma(p(t))$ over that of $\gamma(p(s))$. The increase of the value of $dp^j(t)/dt$ over the value of $dp^j(s)/ds$ is very small.

Conversely, if $\gamma(p(s))$ is close to 1, then the component velocity values are all less than c. The effect of a positive exponent makes little change in $\gamma(p(t))$.[11] Most of the change is due to the increase of $dp(j)(t)/dt$ compared to the value of $dp^j(s)/ds$.

The time component equation, Eq. (6.125), can be integrated to give the time development of $\gamma(p(s))$. To achieve this, one first rewrites the equation as

$$\frac{\gamma(p(s))}{1 - \gamma(p(s))^2} \frac{d\gamma(p(s))}{ds} = cA_0(p(s)). \qquad (6.133)$$

Multiplying by -1 gives

$$\frac{\gamma(p(s))}{\gamma(p(s))^2 - 1} \frac{d\gamma(p(s))}{ds} = -cA_0(p(s)). \qquad (6.134)$$

This equation can be solved by setting $w(p(s)) = \gamma(p(s))^2 - 1$ and integrating. One obtains

$$\ln(\gamma(p(t))^2 - 1) - \ln(\gamma(p(s))^2 - 1) = -2 \int_s^t cA_0(p(u))du. \qquad (6.135)$$

Exponentiation gives

$$\gamma(p(t))^2 - 1 = (\gamma(p(s))^2 - 1) \exp\left(-2 \int_s^t cA_0(p(u))du\right). \qquad (6.136)$$

This equation shows that if the exponent is negative, then for t later than s, $\gamma(p(t))^2 < \gamma(p(s))^2$. Conversely, if the exponent is positive, then $\gamma(p(t))$ is larger than $\gamma(p(s))$.

The integrals in the exponents of Eqs. (6.132) and (6.136) can be evaluated. The relation

$$cA_0(p(u)) = \frac{d\alpha(p_0(u))}{dp^0(u)} \frac{dp^0(u)}{du} = \frac{d\alpha(p_0(u))}{du}, \qquad (6.137)$$

used in the exponent integral, gives

$$\int_s^t cA_0(p(u))du = \int_s^t \frac{d\alpha(p_0(u))}{du}du = \alpha(ct) - \alpha(cs). \qquad (6.138)$$

---

[11]This is the case provided the exponent is not so large that $dp^j(t)/dt$ is close to c.

The space and time geodesic equations (6.132) and (6.136) become

$$\frac{dp^j(t)}{dt}\gamma(p(t)) = \frac{dp^j(s)}{ds}\gamma(p(s))e^{\alpha(cs)-\alpha(ct)} \qquad (6.139)$$

and

$$\gamma(p(t))^2 - 1 = (\gamma(p(s))^2 - 1)e^{2(\alpha(cs)-\alpha(ct))}. \qquad (6.140)$$

The equations for the path q are obtained from the above by replacing p with q. As has been seen, the equations are the same for both paths as they should be. The reason for giving the equations for the path p is that they are easier to read. For example, the velocity of light is represented by c. For the path q, it appears in the equations as $b(q(s))$ where

$$b(q(s)) = e^{-\alpha(x)+\alpha(p(s))}c. \qquad (6.141)$$

The equations for p and q are the same because the various exponential factors appearing in the components of the terms in the converted equation cancel out. Equation (6.114) is an example. Also from Eq. (6.112), one sees that

$$b(q(s))A_\mu(q(s)) = cA_\mu(p(s)).$$

Also the dynamical equations for the gamma factors are the same because $\gamma(p(s)) = \gamma(q(s))$. The resulting equations are

$$\frac{dq^j(t)}{dt}\gamma(q(t)) = \frac{dq^j(s)}{ds}\gamma(q(s))e^{\alpha(cs)-\alpha(ct)} \qquad (6.142)$$

and

$$\gamma(q(t))^2 - 1 = (\gamma(q(s))^2 - 1)e^{2(\alpha(cs)-\alpha(ct))}. \qquad (6.143)$$

These equations show that if $\alpha(cs) > \alpha(ct)$ with t later than s, then $\gamma(q(t)) > \gamma(q(s))$. This corresponds to a contraction of the space at t relative to the space at s. If $\alpha(cs) < \alpha(ct)$, then the space at t is an expansion of the space at s. Note that if α depends on time only, as is the case here, then the local spaces $\mathbb{S}_y^{g(y)}$ for all space locations, $\mathbf{y}$, where $y = \mathbf{y}, cs$ can be identified as the space, $\mathbb{S}_{cs}^{g(cs)}$. The connection between the local spaces $\mathbb{S}_{\mathbf{y},cs}^{g(\mathbf{y},cs)}$ and $\mathbb{S}_{\mathbf{z},cs}^{g(\mathbf{z},cs)}$ is the identity map.

6.5.2.3   *Dependence of path lengths on path location*

A general property of path length or distances between points is that they can depend on the location of the original path p or path points in the base space–time M. The dependence is a function of the space and time variation of the α field and its gradient vector field, $\vec{A}$.

To understand this, let p be a timelike path or world line on M. In the usual global mathematical framework, this path can be described as a bundle of tangent vectors $\vec{p}(s)$. Here s is the time variable. In the local mathematical framework and fiber bundle background used here as $\mathfrak{MF}^g$, the path tangent vector bundle is treated as a section on the part of the fiber bundle with base space of the path points in M. For each s, $\vec{p}(s)$ is lifted to a vector $[\vec{p}(s)]_{g(p(s))}$ in $\mathbb{S}_{p(s)}^{g(p(s))}$ in the bundle fiber at $p(s)$.

Parallel transport of the lifted vector to a vector in a reference space, $\mathbb{S}_x^{g(x)}$, results in the vector $\vec{q}(s)_{g(x)}$ where

$$\vec{q}(s)_{g(x)} = C_g(x, p(s))\vec{p}(s)_{g(x)} = [e^{-\alpha(x)+\alpha(p(s))}\vec{p}(s)]_{g(x)}. \qquad (6.144)$$

This is a repetition of Eq. (6.89). As Eq. (6.92) shows, the path length between points, $p(s)$ and $p(t)$ in $\mathbb{S}_x^{g(x)}$, is given by

$$[c\Delta \tau_s^t]_{g(x)} = \left[e^{-\alpha(x)} \int_s^t e^{\alpha(p(w))} |\vec{p}(w)| dw \right]_{g(x)}. \qquad (6.145)$$

Let p' be the path in M resulting from a space and time translation of the points of p by a fixed quantity, δ, where the μ component of δ is δ^μ. For each time w, define the time w' by

$$w' = w + \delta^0. \qquad (6.146)$$

Here δ^0 is the time component of δ. The relation between p and p' is given by

$$p'(w') = p(w) + \delta. \qquad (6.147)$$

Other than the translation, the two paths are the same. For example, the tangent vectors are the same in that $\vec{p}'(w') = \vec{p}(w)$. If p is the world line of the motion of a particle, then p' is the world line of the same particle moving under the same dynamical conditions. The only change is when and where the motion is occurring.

The path q' is defined from p' the same way as q is defined from p. One obtains

$$\vec{q'}(w')_{g(x)} = C_g(x, p'(w'))\vec{p'}(w')_{g(x)} = [e^{-\alpha(x)+\alpha(p'(s))}\vec{p'}(w')]_{g(x)}$$

$$= [e^{-\alpha(x)+\alpha(p'(w'))}(\vec{p}(w) + \delta)]_{g(x)}. \tag{6.148}$$

The length of q' from time $s' = s + \delta^0$ to $t' = t + \delta^0$ is given by

$$[\Delta\tau_{s'}^{t'}]_{g(x)} = \left[e^{-\alpha(x)} \int_{s'}^{t'} e^{\alpha(p'(w'))} \left(\eta_{\mu,\nu} \frac{dp'^{\mu}(w')}{dw'} \frac{dp'^{\nu}(w')}{dw'} \right)^{1/2} dw' \right]_{g(x)}.$$

$$\tag{6.149}$$

The relation of the length of q' to the length of q can be seen by use of Eqs. (6.147), (6.146), and (6.147) and

$$\frac{dp'^{\mu}(w')}{dw'} = \frac{dp^{\mu}(w)}{dw} \tag{6.150}$$

to express the length of q' by

$$\Delta\tau'_{g(x)} = \left[e^{-\alpha(x)} \int_{s}^{t} e^{\alpha(p(w)+\delta)} \left(\eta_{\mu,\nu} \frac{dp^{\mu}(w)}{dw} \frac{dp^{\nu}(w)}{dw} \right)^{1/2} dw \right]_{g(x)}.$$

$$\tag{6.151}$$

This differs from the length of q by the presence of the δ term in the exponent, $\alpha(p(w) + \delta)$.

The relation between the tangent vector values of the four paths, p, p', q, and q', is illustrated graphically in Figure 6.16. Tangent vectors for the initial and endpoints are shown in the base space, $\mathbb{S}^4 = M$, in the four local spaces for the lifted tangent vectors, and in a reference space, $\mathbb{S}_x^{g(x)}$, at location x in $\mathbb{S}^4$.

6.5.2.4 *Geodesic equation dependence on path location*

In this section, the value subscript $g(x)$ is not shown. It is to be understood that all the quantities and equations refer to values or meanings of quantities in $\mathbb{S}_x^{g(x)}$.

The space–time translation of paths in M also affects the geodesic equations for the local paths in $\mathbb{S}_x^{4,g(x)}$. This can be seen by noting

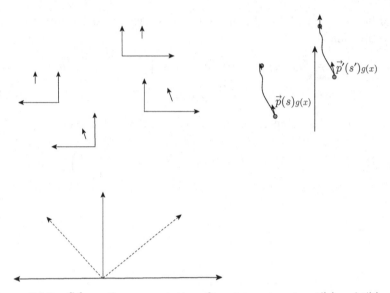

Figure 6.16. Schematic representation of two tangent vectors $\vec{p}(s)$ and $\vec{p}(t)$, and $\vec{p}'(s')$ and $\vec{p}'(t')$ for paths p and p' in space–time $\mathbb{S}^4$, and their lifts and parallel transport to tangent vectors in a reference local space–time $\mathbb{S}_x^{g(x)}$ at location x in $\mathbb{S}^4$. The tangent vector spaces sit at the beginnings and ends of the paths p and p' in $\mathbb{S}^4$. The space–time locations of these four tangent vectors are shown along with the corresponding lifted vectors. The lifts of the four vectors and of the paths to the reference space–time at x are implemented by charts. To avoid even more clutter in the figure, these charts are not shown. The horizontal dashed lines indicate parallel transport of each of the four lifted vectors to tangent vectors $\vec{q}(s), \vec{q}(t), \vec{q}'(s)$, and $\vec{q}'(t)$ in the reference space–time $\mathbb{S}_x^{4,g(x)}$. The shift in the q and q' vector locations and lengths relative to those of p and p' is done to show the difference between the vector values of the q and q' tangent vectors and the vector values of the p and p' tangent vectors. The q and q' vectors are the same vectors as the lifts of the four tangent vectors in $\mathbb{S}^4$.

that the equations for a path q' are obtained from the equations for q, Eq. (6.113), by replacement of p by p' and s (or w) by w'. The equations for q' are

$$\frac{d}{dw'}\left(\frac{dq'^{\nu}(w')}{dw'}\frac{1}{|\vec{q}'(w')|}\right) = \frac{d}{dw'}\left(\frac{dp'^{\nu}(w')}{dw'}\frac{1}{|\vec{p}'(w')|}\right)$$

$$= \eta^{\nu,\nu}A_{\nu}(p'(w'))|\vec{p}'(w')| - A_{\mu}(p'(w'))\frac{dp'^{\mu}(w')}{dw'}\frac{dq'^{\nu}(w')}{dw'}\frac{1}{|\vec{q}'(w')|}.$$

$$(6.152)$$

This equation has the same functional form as does the equation for q in Eq. (6.113).

The relation between this equation and that for q can be seen by converting Eq. (6.152) to an equivalent equation for q. Use of Eqs. (6.146) and (6.147) and $|\vec{p}'(w')| = |\vec{p}(w)|$ and

$$\frac{dq'^{\mu}(w')}{dw'}\frac{1}{|\vec{q}'(w')|} = \frac{dq^{\mu}(w)}{dw}\frac{1}{|\vec{q}(w)|} \tag{6.153}$$

in[12] Eq. (6.152) gives

$$\frac{d}{dw}\left(\frac{dq^{\nu}(w)}{dw}\frac{1}{|\vec{q}(w)|}\right) = \eta^{\nu,\mu}A_{\mu}(p(w)+\delta)|\vec{p}(w)|$$

$$- A_{\mu}(p(w)+\delta)\frac{dp^{\mu}(w)}{dw}\frac{dq^{\nu}(w)}{dw}\frac{1}{|\vec{q}(w)|}. \tag{6.154}$$

This differs from the geodesic equations for the path q by the presence of the term δ in the argument for the components of the vector field, $\vec{A}$.

The distance between the points, $q(s)$ and $q(t)$ in $\mathbb{S}_x^{g(x)}$, is given by Eq. (6.145) with the path q satisfying the geodesic equation, Eq. (6.113). The distance between $q'(s')$ and $q'(t')$, also in $\mathbb{S}_x^{g(x)}$, is given by Eq. (6.149) where q' satisfies Eq. (6.152). The difference between the two distances follows from the space and time dependence of the vector field, $\vec{A}$, in the geodesics equations for paths q and q'. It is also seen directly from a comparison of the path lengths shown in Eq. (6.149) or (6.151) and Eq. (6.145).

6.5.3 $\vec{A}$-dependent space–time geometry

As was noted before, the form of the geodesic equations for the path q shows that the geometries of local spaces deviate from flatness. The equations for the path q in the local space–times, obtained from

---

[12]This equation follows from the equalities,

$$\frac{dq'^{\mu}(w')}{dw'}\frac{1}{|\vec{q}'(w')|} = \frac{dp'^{\mu}(w')}{dw'}\frac{1}{|\vec{p}'(w')|} = \frac{dp^{\mu}(w)}{dw}\frac{1}{|\vec{p}(w)|} = \frac{dq^{\mu}(w)}{dw}\frac{1}{|\vec{q}(w)|}.$$

the equations for the path p in M, have the same functional form as the equations for p. The equations for the path p are not geodesic equations for M as they were derived under the restrictions of lift of tangent vectors followed by parallel transport to a local space.

Another way to describe the geometries of the local spaces is by the form of the metric tensors for the spaces. Let $d\vec{-}$ denote the field of differential vectors on M. For each location, y, $d\vec{y}$ is the differential vector at y. Let $d\vec{-}_{g(-)}$ represent the lift of this field as a section on the fiber bundle, $\mathfrak{MF}^g$. For each y, $d\vec{y}_{g(y)}$ is a differential vector with length, $|d\vec{y}|_{g(y)}$, at $y_{g(y)}$ in $\mathbb{S}_y^{g(y)}$.

Expression of geometric properties of extended objects, such as lengths, areas, and volumes of paths, surfaces, and solids, cannot be done using the lifted differential vector field $d\vec{-}_{g(-)}$. The requisite integrals are not defined.

This problem is fixed by parallel transport of the differential vectors, $d\vec{y}_{g(y)}$, to a space–time, $\mathbb{S}_x^{g(x)}$, at a reference location, x. The resulting differential vectors become $[e^{-\alpha(x)+\alpha(y)}d\vec{y}]_{g(x)}$ with vector value $e^{-\alpha(x)+\alpha(y)}d\vec{y}$. The length[13] of this vector is denoted by

$$|e^{-\alpha(x)+\alpha(y)}d\vec{y}| = e^{-\alpha(x)}|e^{\alpha(y)}d\vec{y}|.$$

This is the value of a number in $\bar{R}_x^{g(x)}$.

Define a scaled differential vector $D_\alpha\vec{y}$ by

$$D_\alpha\vec{y} = e^{\alpha(y)}d\vec{y}. \tag{6.155}$$

The corresponding length element is defined by

$$\begin{aligned}
|D_\alpha\vec{y}| &= e^{\alpha(y)}|d\vec{y}| = \{e^{2\alpha(y)}\eta_{\mu,\nu}d^\mu y d^\nu y\}^{1/2} \\
&= \{\beta_{\mu,\nu}(y)d^\mu y d^\nu y\}^{1/2}.
\end{aligned} \tag{6.156}$$

Here $D_\alpha\vec{y}$ is the value of a vector in $\mathbb{S}_y^{g(y)}$ and $|D_\alpha\vec{y}|$ is the value of the number, $[|D_\alpha\vec{y}|]_{g(y)}$, in $\bar{R}_y^{g(y)}$.

The length value, $|D_\alpha\vec{y}|$, shows the location dependence of the distance element in each local space–time. It is not normalized in

---

[13]The distinction between vectors and numbers and their values can be ignored as long as one is working within a single local space, such as $\mathbb{S}_x^{g(x)}$.

that it lacks the condition that for an image space–time at each location x, $|D_\alpha \vec{x}| = |d\vec{x}|$. This can be fixed by defining an image space location-dependent distance element by

$$|D_{x,\alpha}\vec{y}| = e^{-\alpha(x)+\alpha(y)}|d\vec{y}| = \{e^{-2\alpha(x)+2\alpha(y)}\eta_{\mu,\nu}d^\mu y d^\nu y\}^{1/2}$$

$$= \{\beta_{x,\mu,\nu}(y)d^\mu y d^\nu y\}^{1/2}. \tag{6.157}$$

This equation suggests that $\beta_x(y)$ with components, $\beta_{x,\mu,\nu}(y)$, plays the role of a metric tensor for the space–time, $\mathbb{S}_x^{g(x)}$. The dependence of $\beta_x(y)$ on y shows the location dependence of the metric tensor within the image space–time at x. It is normalized to $\mathbb{S}_x^{g(x)}$ in that for the point $x_{g(x)}$ in the local space–times at x, $\beta_x(x) = \eta$.

The normalization of $\beta_x(y)$ is different for different local space–times. For example, $\beta_z(y)$ in $\mathbb{S}_z^{g(z)}$ differs from $\beta_x(y)$ in $\mathbb{S}_x^{g(x)}$ by the factor, $e^{-\alpha(z)+\alpha(x)}$. That is,

$$\beta_z(y) = e^{-\alpha(z)+\alpha(x)}\beta_x(y)$$

independent of y.

Equation (6.157) shows that in the presence of α, the metric tensor changes from that for a flat space–time to

$$\beta_x(y) = e^{-2\alpha(x)+2\alpha(y)}\eta. \tag{6.158}$$

The components of $\beta_x(y)$ are given by the components of η scaled by the exponential factor as in

$$\beta_{x,\mu,\nu}(y) = e^{2(-\alpha(x)+\alpha(y))}\eta_{\mu,\nu}. \tag{6.159}$$

In these definitions, $e^{-2\alpha(x)+2\alpha(y)}$ is the value of $[e^{-2\alpha(x)+2\alpha(y)}]_{g(x)}$ as a number in $\mathbb{S}_x^{g(x)}$.

The difference between $\beta_x(y)$ and η is consistent with the deviations of the geodesic equations, Eqs. (6.117) and (6.118), from those for a straight line. This can be seen from the expression for the length value of a path q in $\mathbb{S}_x^{g(x)}$. The expression is

$$L(q) = \int_s^t |dq(u)|du = \int_s^t \left\{\beta_{x,\mu,\nu}(p(u))\frac{d^\mu p(u)}{du}\frac{dp^\nu(u)}{du}\right\}^{1/2}du. \tag{6.160}$$

This repeats Eq. (6.93) which was used to obtain the geodesic equations.

It has been seen that $\beta_x(y)$ is obtained as a metric tensor for the local space–time at x by treating $d\vec{\,}_{g(-)}$ as a vector section on the fiber bundle $\mathfrak{MF}^g$ followed by parallel transport of the local vectors to $\mathbb{S}_x^{g(x)}$. The same result can be obtained by treating the lengths, $|d\vec{y}|$, of $d\vec{y}$ as a scalar section, $|d\vec{\,}|_{g(-)}$, on the fiber bundle. Parallel transport of the differential lengths, $|d\vec{y}|_{g(x)}$, to the space–time $\mathbb{S}_x^{g(x)}$ results in the local differential length field, $e^{-\alpha(x)+\alpha(y)}|d\vec{y}|_{g(x)}$, as a function of location, y, in $\mathbb{S}_x^{g(x)}$. Here $\beta_x(y)$ is obtained by bringing the exponential factor inside the square root appearing in the definition of the element.[14]

Finally, it is worth noting that in the case of α depending on time only, the comoving distance [73] between two systems is the same for all times. To see this, let O_1 and O_2 be systems at space locations $\mathbf{y}$ and $\mathbf{z}$ at time s in M. The images of the locations of these systems in $\mathbb{S}_{cs}^{g(cs)}$ are the points $[\mathbf{y}, cs]_{cs}$ and $[\mathbf{z}, cs]_{cs}$. Since the metric tensor $\beta_{cs}(cs) = \eta$ for $\mathbb{S}_{cs}^{g(cs)}$, the distance between the two image points is $|\mathbf{z}_{g(cs)} - \mathbf{y}_{g(cs)}| = [|\mathbf{z} - \mathbf{y}|]_{g(cs)}$ at time s.

As time progresses, the point image locations and the distance between the objects flow through the successive space–times. For any time t after s, the systems are at locations $[\mathbf{z}, ct]_{g(ct)}$ and $[\mathbf{y}, ct]_{g(ct)}$ in the space–time $\mathbb{S}_{ct}^{g(ct)}$. Since the metric tensor for this space at the time t is $\beta_{ct}(ct) = \eta$, the distance between the points is $[|\mathbf{y} - \mathbf{z}|]_{ct}$.

Comparison of this distance value with that for the time s shows that the distance value between the two objects is the same at the two times. This shows that the value of the comoving distance between the two objects is unchanged as time progresses.

6.5.4   *Relation between M and the image space–times*

At present, one has arrived at the situation where the local space–times deviate from flatness as described by the metric tensors, $\beta_x(y)$, for each point x in M. If M is a flat space–time, then the local space–times are not faithful images of M.

---

[14]It is tempting to speculate here that the effect of the α field and its gradient on the space–time metric tensor extends to the metric tensor, $g_{\mu,\nu}$, of general relativity. Whether such an extension makes sense or not requires more work.

One would like the geometry of M to be such that the local space–times are faithful or chart images of M. However, this is problematic. The reason is that the geometries of the local space–times differ by a scale factor based on the locations of the space–times in M.

This difference is shown by the location dependence of the metric tensors for the different space–times. The metric tensors for the points $y_{g(x)}$ and $y_{g(z)}$ in $\mathbb{S}_x^{g(x)}$ and $\mathbb{S}_z^{g(z)}$ are $\beta_x(y)$ and $\beta_z(y)$. The distance element, $|d\vec{y}|$, in M becomes

$$|\beta_x(y)d\vec{y}| = e^{-\alpha(x)}(e^{2\alpha(y)}\eta_{\mu,\nu}d^\mu(y)d^\nu(y))^{1/2}$$

in the space–time at x and

$$|\beta_z(y)d\vec{y}| = e^{-\alpha(z)}(e^{2\alpha(y)}\eta_{\mu,\nu}d^\mu(y)d^\nu(y))^{1/2}$$

in the local space at z.

It is possible to alter the geometry of M by inclusion of the α factor. In this case, for each point y in M, the differential distance element for each point y would become $e^{\alpha(y)}|d\vec{y}|$. However, this does not take account of the factors $e^{-\alpha(x)}$ and $e^{-\alpha(z)}$ for the spaces at x and z.

It turns out that the location dependence and deviation of the geometries of the local space–times from flatness are not as problematic as they seem. The reason is that the restrictions on the space–time variation of the value field, α, are such that the differences between the geometries of the image space–times and M are far too small to be observed *locally*. The meaning of this is that

$$\beta_x(y) \approx \beta_z(y) \tag{6.161}$$

for all local locations x and z in M.

It will be seen later on that setting the geometry of M to be that of $\beta_x(y)$ for all local x and all y resolves the problem of the relation between the local space–times and M. For local points, x, the geometries of $\mathbb{S}_x^{g(x)}$ and M are essentially the same. If $\beta_M(y)$ is the metric tensor for M, then

$$\beta_x(y) \approx \beta_z(y) \approx \beta_M(y) \tag{6.162}$$

for all local x and z and all y.

If α is constant everywhere in space–time, the geometries of M and the local space–times are the same. In this case, the metric tensor for the local geometries is η.

It should be noted that M is the space–time in which events happen. Physical systems move and interact with one another. Observers are also physical systems in M. The motion of an observer traces out a world line in M. At each point of the world line for an observer, the local mathematics available to an observer is that collocated with the observer. This is the mathematics used by an observer to create theoretical descriptions of physical systems. The geometries for the theoretical descriptions are those of the space–time regions around each point of the world line for the observer. Local space–times and local mathematics provide the arena in which the descriptions obtain meaning for a collocated observer.

6.5.5 *The effect of the value field on light*

The description of the effect of the value field on path lengths presented so far does not apply to light. The reason is that all lightlike paths have zero length. Yet one knows from physics that two events at different space–time locations are separated by a distance measured by the time it takes light to travel from the event at an earlier time to the event at a later time. Typically, these distances are measured in light time units, such as light seconds, light minutes, and light years.

An examination of the effect of the value field, α, on light or electromagnetic radiation in general is limited to the parameters that can vary. The speed value, c, of light is not a parameter that can vary. This holds true in the framework of local mathematics with space–times at each location in the base space–time. For each location, y in M, the speed of light in the local space–time, $\mathbb{S}_y^{g(y)}$, is $c_{g(y)}$. The numerical quantities, $c_{g(y)}$ and $c_{g(z)}$, are different. They both have the same value, c.

The wavelength of light can be defined directly from the wave properties of light. Let s be the arrival time of a wave crest at space location, $\mathbf{u}$. The time $s + \Delta s$ is defined to be the arrival time of the next wave crest at $\mathbf{u}$. The wavelength, γ, is the distance between successive wave crests. It is related to time interval between crests by $\gamma = c\Delta s$.

Let $z = ct, \mathbf{z}$ be a reference location in M. Light emitted by an event at any location on the past light cone from z will be first visible at time t and location $\mathbf{z}$. Let $u = cs, \mathbf{u}$ be a z light cone location of an event emitting light. The time s is earlier than t.

This is shown in Figure 6.17. For ease in presentation, the figure is shown for two space dimensions and one time dimension. The light path, p, from $s, \mathbf{u}$ to the observer at $t, \mathbf{z}$ is shown as a dashed line.

The motion of light from u from u to z along the light cone path in M can be represented by

$$(cds)^2 - (d\mathbf{u} \cdot d\mathbf{u}) = 0 \tag{6.163}$$

or

$$cds = (d\mathbf{u} \cdot d\mathbf{u})^{1/2}. \tag{6.164}$$

Replacement of ds by Δs and $d\mathbf{u}$ by $\delta\mathbf{u}$ gives $(\delta\mathbf{u} \cdot \delta\mathbf{u})^{1/2}$ equal to the wavelength.

An observer at z would like to experimentally compare the wavelength of the arriving light with light emitted locally from a source of the same type as the emitting source at u. Possible outcomes are that the wavelength of the local light can be the same, longer, or shorter than that of the arriving light.

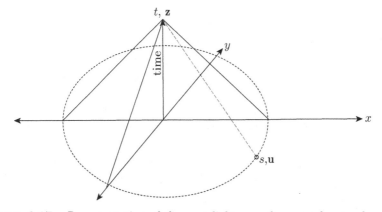

Figure 6.17. Representation of the past light cone from an observer location in space–time at $t, \mathbf{z}$. The circle at time s gives the space locations of all events that would be visible at the observer's location. The light path for an event at $s, \mathbf{u}$ to the observer is shown as a dashed line. For ease in presentation, space is shown as two-dimensional.

The usual theoretical description of the wavelength of light as it moves along the light cone path can be obtained from Eq. (6.164). If $(\delta \mathbf{u} \cdot \delta \mathbf{u})^{1/2}$ is the wavelength at u, then $(\delta \mathbf{z} \cdot \delta \mathbf{z})^{1/2}$ is the wavelength of light upon arrival at z.

The arena of local mathematics and the presence of the value field, α, require a change in this description. The reason is that the wavelength of the light is affected by variations in the α field along the light cone path. These variations are described by the vector field, $\vec{A}$, as the gradient field of α.

Let $z = \mathbf{z}, ct$ be the location in the universe of an observer in M. This location also is the location of us as observers in the cosmological universe. The time t is the age of the universe, about 14 billion years.

Events that are electromagnetically visible to the observer lie on the past light cone from z. Electromagnetic radiation emitted by an event at z would become visible to an observer on the future light cone from z.

The local space–time and real number structure at z in the base space–time, M, are $\mathbb{S}_z^{g(z)}$ and $\bar{R}_z^{g(z)}$. For the point u, the local space–time and real number structures are $\mathbb{S}_u^{g(u)}$ and $\bar{R}_u^{g(u)}$.

The observer knows that the wavelength of light emitted from a source at u is a number in $\bar{R}_u^{g(u)}$. Also known is the fact that the wavelength of light emitted from a local source at z is represented by a number in $\bar{R}_z^{g(z)}$. If the sources are of the same type, then experimental comparison of the wavelength values of the light arriving from u with the wavelength of light emitted locally at z would determine the difference, if any, between the wavelength values of the light from the distant source with that emitted locally. If there is an experimental difference in the wavelength values, one would like to be able to give a theoretical description of the difference that agrees with experiment.

In the arena of local mathematical structures at different locations, it is not possible to give a direct theoretical description of the comparison of the wavelength value of light as a number value in $\bar{R}_u^{g(u)}$ with the value of the wavelength of light as a number in a number structure, $\bar{R}_z^{g(z)}$, at location z. As has been seen before, this problem is fixed by parallel transport of the wavelength of light from the distant source at u to z. This enables a theoretical description of differences in the wavelength values locally within the same real number structure.

Assume that an event at u in M emits light with wavelength value γ. The local description of this event is as an event at point $u_{g(u)}$ in $\mathbb{S}_u^{g(u)}$. The wavelength of the emitted light is the number, $\gamma_{g(u)}$, in $\bar{R}_u^{g(u)}$ with value γ.

Parallel transport of this numerical quantity to an observer at z is implemented by a number preserving number value changing connection, $C_g(z, u)$. Specifically, one has

$$\lambda(u)_{g(z)} = C_g(z, u)\gamma_{g(u)} = [e^{-\alpha(z)+\alpha(u)}\gamma]_{g(z)}. \tag{6.165}$$

For an observer at z, this equation shows that the wavelength, $\lambda(u)_{g(z)}$, of the transported light, as a numerical quantity in $\bar{R}_z^{g(z)}$, is the same as the wavelength, $\gamma_{g(u)}$, of light emitted at the source, u. But the wavelength values of these two numbers differ by the exponential factor.

The wavelength value of the light emitted at u and arriving at z is greater than the wavelength value of light emitted at z if the exponent value is positive. If the exponent is negative, the wavelength value of light arriving at z from u is less than that emitted locally at z. Note that $\gamma_{g(z)}$, as the wavelength of locally emitted light, has the same wavelength value as does $\gamma_{g(u)}$ for the light emitted at u. However, $\gamma_{g(z)}$ is not the same number as is $\gamma_{g(u)}$.

As has been emphasized in previous sections and chapters, one is interested in the values or meanings of numerical quantities. Theoretical descriptions are meaningful. They are represented by values of mathematical elements, such as the values of vectors or numbers. For this reason, the theoretical description of the comparison of light arriving from u to light at z consists of a comparison of wavelength values, not wavelengths. The difference in the wavelength values is a theoretical prediction to be supported or refuted by experiment.[15]

---

[15]Recall that a numerical quantity such as a wavelength has no meaning or value by itself. It acquires meaning or value to an observer at z only as a number within a real number structure, $\bar{R}_z^{g(z)}$. The structure containing the numerical quantity provides the environment in which the number acquires meaning or value. The *values* of the wavelengths and energies have meaning to an observer at z. The wavelengths and energies as numbers are unchanged by the parallel transport.

Figure 6.18 shows the lift of the wavelength of light at $u = \mathbf{u}, s$ in the base space–time to the local space–time at u. This is followed by parallel transport to the local space–time at z. The half oval at u on the future light cone denotes the locations of all events at time s that are visible to an observer at $z = t, \mathbf{z}$. To save on space, only half of the event oval is shown at time, s. The local space–times, $\mathbb{S}_u^{g(u)}$ and $\mathbb{S}_z^{g(z)}$, and real number structures, $\bar{R}^{g(u)}$ and $\bar{R}_z^{g(z)}$, at u and z are shown above the points, u and z, in the base space–time.

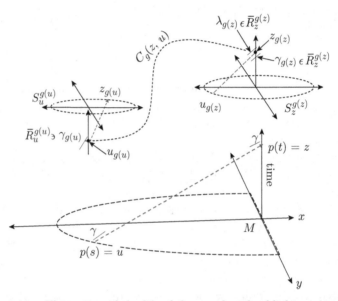

Figure 6.18. Illustration of the lift of the wavelength of light, emitted by an event at u on a light cone path in the base space–time, to a collocated local space–time, $\mathbb{S}_u^{g(u)}$, and parallel transport to a wavelength in a local reference space–time at the light cone apex, z. Wavelengths are shown as short line segments. The real number structures containing the wavelengths at z and u are shown. Parallel transport of the wavelength at s to z is shown as a dashed curve. The wavelength of the light emitted locally at z from the same source type as occurred at u is also shown. The light paths in each space–time are shown as dashed straight lines. These are along the future light cone from $u_{g(u)}$ in $\mathbb{S}_u^{g(u)}$ and the past light cone from $z_{g(z)}$ in $\mathbb{S}_z^{g(z)}$. The points $z_{g(z)}$ and $u_{g(z)}$ in $\mathbb{S}_u^{g(u)}$ and $\mathbb{S}_z^{g(z)}$ represent respectively the future light reception point z and past emission point u. The wavelength of light at both sites u and z in the base space is shown as γ with no subscript. The local space–times, $\mathbb{S}_u^{g(u)}$ and $\mathbb{S}_z^{g(z)}$, are shown directly above their locations in M. Only one half of the base space–time and past event oval are shown. Space is shown as two-dimensional.

The wavelengths are shown by numerical quantities next to short line segments. $\lambda_{g(z)}$ denotes the parallel transport of the wavelength $\gamma_{g(u)}$ to the local space–time and real number structure at z. Space is shown as two-dimensional with time the vertical axis.

6.5.6 α and space expansion or contraction

So far, the effect of the presence of the α field on the wavelength has been described in mathematical terms as the effect of parallel transport of a wavelength from a local space at one location to a local space at another location.

The effect can also be given a more physical description by considering the motion of light along a light cone path in M. In the arena of local mathematics, the motion of light is described theoretically as passing through a sequence of local space–times located at each point on the light cone. At each point u on the light cone, the light is described theoretically by a wavelength in the local space–time, $\mathbb{S}_u^{g(u)}$, at u.

The effect of the α field on the wavelength is a consequence of the effect of the field on the size or extent of the coordinate part of the local space–times on the past light cone from the observer location at z. The relative sizes of the spaces are determined by the α-dependent scaling factor in the metrics of the local space–times.

For each point u, the metric for the space–time, $\mathbb{S}_u^{g(u)}$, at u is given by

$$\beta_u = e^{-2\alpha(u)}\eta. \tag{6.166}$$

The length of a short path at u in the local space is given by

$$d\tau_u = (e^{-2\alpha(u)}\eta_{\mu,\nu}d^\mu u d^\nu u)^{1/2} = (e^{-2\alpha(u)}(cds^2 - (d^j u)^2)))^{1/2}. \tag{6.167}$$

If $d\tau_u$ is a differential along a light cone path, then $d\tau_u = 0$. This gives

$$e^{-2\alpha(u)}cds^2 = e^{-2\alpha(u)}(d^j u)^2. \tag{6.168}$$

Summation over repeated indices is implied. Also $u = \mathbf{u}, cs$.

These results are used to compare relative sizes of spaces along a light cone path. For the reference location, $z = \mathbf{z}, ct$, the spatial extent of the space at a point u on either the past or future light cone from z is obtained by adding a factor $e^{2\alpha(z)}$ to the metric. Equation (6.168) becomes

$$e^{2\alpha(z)-2\alpha(u)}cds^2 = e^{2\alpha(z)-2\alpha(u)}(d^ju)^2. \qquad (6.169)$$

The presence of $\alpha(z)$ in the exponent relativizes the spatial size at u to that of the space at z. It shows that as one moves along the past light cone to z, the size of the space at u approaches that of the space at z. At the point $u = z$, the exponent is zero and the metric becomes $\beta_z(z) = \eta$. This shows that the two spaces are the same size and are identical.[16]

The relative sizes of the spaces can be expressed in terms of space expansion and contraction. If the exponent in Eq. (6.169) is negative, it means that the space at u is a contraction of the space at z. If the exponent is positive, the space, $\mathbb{S}_u^{g(u)}$, is an expansion of the space $\mathbb{S}_z^{g(z)}$.

Let λ_u be the wavelength value of light at u and λ_v the wavelength value of the same light at another point v on the past light cone from z. At each point on the light cone, the scaling factors for the wavelength values are the same as those for the metrics. This can be expressed by

$$\frac{\lambda_u}{e^{-\alpha(u)+\alpha(z)}} = \frac{\lambda_v}{e^{-\alpha(v)+\alpha(z)}}. \qquad (6.170)$$

Cancellation of the common exponential factors and rewriting gives

$$\lambda_v = e^{-\alpha(v)+\alpha(u)}\lambda_u. \qquad (6.171)$$

Note that λ_u and λ_v are the wavelength values of the wavelengths $(\lambda_u)_{g(u)}$ and $(\lambda_v)_{g(v)}$.

The effect of the s dependence of α on the wavelength of radiation as it moves along a light cone path in the space–time, M, is shown in Figure 6.19. The wavelength of light emitted at a point u is shown

---

[16]The same description applies to the temporal extent of the space at u. This is seen from $du = d\mathbf{u}, cds$. At time s, the temporal extent ds relative to dt is given by

$$ds = e^{-\alpha(u)+\alpha(z)}dt.$$

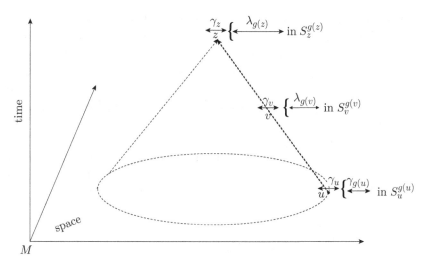

Figure 6.19. Comparison of the wavelength values of light emitted from sources of the same type at $u, v,$ and z and wavelengths of light emitted at u and arriving at v and at z along the light cone in M. Wavelength vectors for emitted light at each of the points are shown by two headed arrows labeled $\gamma_u, \gamma_v,$ and γ_z. The length values of these vectors are all the same. This reflects the fact that the three numbers, $\gamma_u, \gamma_v,$ and γ_z are different numbers. They all have the same number value, γ, in their respective number structures. The wavelength vectors for light emitted at u and moving to v and z are also shown as two headed vectors. The figure shows that the vector lengths, as the same numbers in their respective structures, have different values. The lengths of these vectors also illustrate the expansion of the local space–times as time progresses.

at u and at a transit point, $v = cw, \mathbf{v}$ at time, $w > s$, on to the point, $z = ct, \mathbf{z}$. The wavelengths of light emitted locally at v and z are shown for comparison. The local space–times containing these wavelengths are also shown. The red-shifted wavelengths at v and z are denoted by $\lambda_{g(v)}$ and $\lambda_{g(z)}$. The shifted wavelengths are related to the wavelengths of the locally emitted light by the usual relation:

$$\lambda_{g(v)} = [e^{-\alpha(v)+\alpha(u)}\gamma]_{g(v)}. \qquad (6.172)$$

Equation (6.171) gives the effect of the space scaling factor on the wavelength values of light as it moves along a light cone in M. It shows that the scaling factor has the same effect on the wavelength as it does on the sizes of the local spaces on the light cone path. The extent of a space is mirrored by the length value of the wavelength of light in the space.

Comparison of Eq. (6.171) with Eq. (6.165) shows that the description of the effect of the α field on the wavelength value of light as it passes through local spaces of different sizes is the same as parallel transport of the wavelength from one location on the light cone to another. This adds support to the observation that parallel transport as a purely mathematical operation has physical content.

The effects of variations in the α field on the wavelength of light have experimental consequences. An observer at z can compare the wavelength or energy of light arriving from a source at u with the wavelength of light of the same source type emitted from a local source. The Lyman alpha line in the spectrum of hydrogen is an example.

The amount of red shift from u to z can be defined in the same manner as is done in cosmology. The amount is defined by[17] [74]

$$w_z(u) = \frac{\lambda_{g(z)} - \gamma_{g(z)}}{\gamma_{g(z)}} = \left[\frac{\lambda - \gamma}{\gamma}\right]_{g(z)} = [e^{-\alpha(z)+\alpha(u)} - 1]_{g(z)}.$$

$$(6.173)$$

Here the exponent is positive. If the exponent $-\alpha(z) + \alpha(u)$ is negative, the amount of blue shift is given by

$$\frac{\gamma_{g(z)} - \lambda_{g(z)}}{\lambda_{g(z)}} = \left[\frac{\gamma - \lambda}{\lambda}\right]_{g(z)} = [e^{\alpha(z)-\alpha(u)} - 1]_{g(z)}. \qquad (6.174)$$

6.6 α restricted to depend on time only

For the following discussion, it is useful to restrict α to depend on time only and not on space. Then $\alpha(u) = \alpha(cs)$ where $u = cs, \mathbf{u}$ and $\alpha(z) = \alpha(ct)$ where $z = ct, \mathbf{z}$. The time t is the present cosmological time from the big bang, and s varies from the big bang time, 0 to t.

For wavelength parallel transport,

$$\lambda = e^{-\alpha(ct)+\alpha(cs)}\gamma; \quad \mu = \frac{c}{\lambda} = e^{\alpha(ct)-\alpha(cs)}\frac{c}{\gamma}. \qquad (6.175)$$

The second equation shows the effect of wavelength parallel transport on the frequency. The corresponding energy value of the light at z is

---

[17] w is used to denote the red shift to avoid conflict with z as a location parameter.

given by

$$E(t) = \frac{hc}{\lambda} = \frac{hc}{\gamma}e^{\alpha(ct)-\alpha(cs)} = E(s)e^{\alpha(ct)-\alpha(cs)}. \tag{6.176}$$

Here $E(s)$ is the energy of the light emitted at u.

This result shows that, if $\alpha(cs) > \alpha(ct)$, the energy of light at time t from a distant source at time, s, is less than the energy of light from the same source type emitted locally at ct. This is the expected consequence of the lengthening of the wavelength of light traveling from u to z.

Equations (6.175) and (6.176) are for wavelength values, energy values, and frequency values. The corresponding equations for wavelengths and frequencies would have the subscript $g(z)$ attached.

Application of the description of the effect of variation of $\alpha(cs)$ on wavelengths to cosmology requires that one be clear about the framework for these equations. The framework for this description consists of local mathematics, local space–times, and the presence of a value field, α. Within this framework, the equations describe the meanings or values of the red shift or blue shift as quantities resulting from experimental measurements of the shift of the spectrum of light emitted from different sources at different distances. The base space–time, M, serves as a location frame for events occurring in the past or future. The local space–time at z in M is $\mathbb{S}_z^{g(z)}$.

6.6.1 α and the cosmological red shift

The fact that the time variation of α can describe red shifts in general suggests that it can account for the red shift as observed in cosmology. The restriction of α to depend on time only means that if $u = cs, \mathbf{u}$ and $z = ct, \mathbf{z}$, then

$$\alpha(u) = \alpha(cs) \quad \text{and} \quad \alpha(z) = \alpha(ct). \tag{6.177}$$

The location $z = ct, \mathbf{z}$ is our location in the universe. Locations $u = cs, \mathbf{u}$ are any space points at time s on the past light cone from z. These points are all at a distance $c(t - s)$ from z.

Variations in the field, $\alpha(cs)$, can be used to describe the cosmological red shift of light. Equation (6.173) shows that the red shift can be expressed in the form [74]

$$[w_t(s)]_{g(ct)} = \left[\frac{\lambda_t(s)}{\gamma_s} - 1\right]_{g(ct)} = [e^{\alpha(cs)-\alpha(ct)} - 1]_{g(ct)}. \quad (6.178)$$

This equation gives the red shift of light emitted at time s as seen by an observer at location $z = ct, \mathbf{z}$.

At this point, the functional dependence of α on time is not known. However, the time dependence of α can be such as to account for the fact that the red shift of light increases as the distance between the observer at z and the source at cs increases. Equation (6.178) shows that this happens if $\alpha(cs) > \alpha(ct)$ and $\alpha(cs)$ decreases as s increases. The slope

$$d\alpha(cs)/ds = cA(cs) \quad (6.179)$$

is negative. That is, $A(cs) < 0$.

For times s for which the red shift is small, Eq. (6.178) becomes

$$[w_t(s)]_{g(ct)} \approx [\alpha(cs) - \alpha(ct)]_{g(ct)}. \quad (6.180)$$

In cosmology, it is known that the rate of red shift increase with increasing distance or decreasing time, s, is given by the Hubble parameter, $H(t)$ [75]. The relation of H_0 to the red shift is given by

$$H_0 = -\frac{d}{ds}\frac{\lambda - \gamma}{\gamma} = -\frac{dw_t(s)}{ds}$$

$$= -\frac{d}{ds}(e^{-\alpha(ct)+\alpha(cs)} - 1) = -cA(cs)e^{-\alpha(ct)+\alpha(cs)}. \quad (6.181)$$

The fact that H_0, as a measure of the red shift, is very small means that there is a range of times, s, for which the red shift is very small. For these times, $\alpha(ct) \approx \alpha(cs)$ and the exponential is the identity. The relation between the Hubble parameter and $A(cs)$ is given by

$$-cA(cs) = cA = H_0. \quad (6.182)$$

Here $H_0/c = A$ is the constant value of $-A(cs)$ for times where the red shift is small. Since $\alpha(cs) = \alpha(ct - \delta)$, one has

$$-cA(ct - \delta) = H_0. \quad (6.183)$$

The validity of Eq. (6.182) for a large range of times is based on the fact that the Hubble parameter is very small locally. This is shown by a change of the length units. The value of H_0 is approximately 70 km/sec/Mpc (70 kilometers per second per megaparsec) [75]. A megaparsec is 3.26 million lightyears. In units of inverse years, the Hubble parameter value is

$$H_0 \approx \frac{70\,\text{km/s}}{3.26 \times 10^6 \,\text{years} \times 3 \times 10^5 \,\text{km/s}} \approx 7.2 \times 10^{-11} \,\text{year}^{-1}.$$

(6.184)

In units of inverse seconds,

$$H_0 = 2.3 \times 10^{-18}\,\text{s}^{-1}.$$

(6.185)

This representation of the Hubble constant in inverse time units is useful in that the constant is independent of the distance units. For example, $H_0 = 7 \times 10^{-11}\,\text{km/yr/km} = 7 \times 10^{-11}\,\text{cm/yr/cm}$. This result also shows how small cA is or how small the variation of $-\alpha(ct) + \alpha(cs)$ is with time, s. One sees that the rate of increase of $-\alpha(ct) + \alpha(cs)$ as s decreases from t is extremely slow.

Experiment shows that, for times where the red shift is less than 1, the rate of red shift increase with increasing distance begins to accelerate from the value given by the Hubble parameter. The acceleration begins about 4 billion years in the past [76]. This change can be taken into account by the α field by requiring that the gradient field $\vec{A}(cs)$ increases as s increases.

This acceleration of the rate of red shift is shown by an increase in the downward slope of α as s increases. The change is given by the second derivative of $e^{-\alpha(ct)+\alpha(cs)}$ as

$$\frac{d^2 e^{-\alpha(ct)+\alpha(cs)}}{d(t-s)^2} = \frac{d^2 e^{-\alpha(ct)+\alpha(cs)}}{ds^2} = c \frac{d(A(cs)e^{-\alpha(ct)+\alpha(cs)})}{ds}.$$

(6.186)

If this derivative is negative, it implies an increase in the rate of red shift as s increases. If the derivative is positive, it implies a decrease of the rate of red shift as s increases.

The increase in the rate of red shift as time increases has been described as the main evidence for the existence of dark energy [77]. It follows that the functional dependence of α on time can be such as to account for both Hubble expansion and the effect of dark energy.

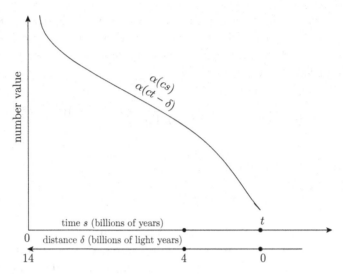

Figure 6.20. A possible dependence of the value field, α, on cosmological time and distance. The time, s, ranges from 0, the time of the big bang, to the present time, t. The distance extends from 0 at our location back to the big bang at 14 billion light years. The upturn of the curve for small s values is followed by a region with a constant negative slope. This is followed by a downturn starting about 4 billion years ago.

The time dependence of α that accounts for the red shift from Hubble expansion and acceleration due to dark energy is represented schematically in Figure 6.20. This is shown by a region of constant negative slope. A downturn in the slope begins about 4 billion years ago and extends to the present. Two abscissa are shown. One is for the time from the big bang at time 0 to the present. The other is for the distance from the present to the 14 billion light year distance of the big bang.

The figure shows that the time derivative or slope of α is negative. The distance derivative has the opposite sign as it is positive. These two representations are given by[18]

$$\frac{d\alpha(cs)}{ds} = -cA(cs) \quad \text{and} \quad \frac{d\alpha(\delta)}{d\delta} = A(\delta). \tag{6.187}$$

---

[18] For convenience, the sign of the vector field time component in this equation is the opposite of that in Eq. (6.96), which was used to derive the geodesic equations.

Here $\delta = c(t - s)$ is the distance from us to the light source. The factors $A(cs)$ and $A(\delta)$ are positive and $A(cs) = A(\delta)$.

The upturn in the curve in the figure for small s has been drawn to show the value of the α field increasing rapidly as s approaches 0. In the limit where $s = 0$, the time of the big bang, $\alpha(0) = \infty$. From Eq. (6.176), one sees that in the limit of $s = 0$, the energy of matter becomes infinite.

This description of the possible effects of the α value field and its derivatives shows that the dependence of the value field on s *can be* such that $cA(cs)$ and $-cdA(cs)/ds$ give a theoretical description of the red shift effects of the Hubble recession and dark energy that agree with experimental results. The upturn at small times with $\alpha(cs) \to \infty$ as $t \to 0$ suggests that α may also be able to account for the big bang. In the limit $s = 0$, the local space has zero extent. All wavelengths shrink to 0. Energies of all systems become infinite.

So far, it has been shown that the time dependence of α can be such to describe the Hubble expansion and the acceleration due to dark energy. It may even be able to account for the big bang. Whether α must have these properties or is unrelated to cosmological effects is not known. More work is needed to give support for (or refutation of) these effects of the α field and its derivatives.

6.6.2 α and space expansion and contraction

So far, it has been seen that the geometry of the local space–times differs from that of M in that a time-dependent scale factor is present. It was also seen that the dependence of α can be such that it describes the red shift due to the Hubble expansion and possibly dark energy.

It is known that the cosmological red shift is caused by an expansion of space. The expansion of space can be shown to be a consequence of the time dependence of the value field shown in Figure 6.20. If α depends on time, only then the expansion or contraction of space is isotropic.

The α-dependent expansion of space can be seen by reference to the metric tensors of earlier local space–times as seen in later local space–times. Let $x = ct, \mathbf{x}$ be our location in M and $y = cs, \mathbf{y}$ be another location in M. The local space–times at these locations are $\mathbb{S}_x^{g(x)}$ and $\mathbb{S}_y^{g(y)}$. If α depends on time only, as in $\alpha(cs)$, then the metric tensors at points $x_{g(x)}$ in $\mathbb{S}_x^{g(x)}$ and $y_{g(y)}$ in $\mathbb{S}_y^{g(y)}$ are given by

$$\beta_{ct}(x) = \eta_{g(x)} \quad \text{and} \quad \beta_{cs}(y) = \eta_{g(y)}. \tag{6.188}$$

Let y be on the past light cone of x. Then t is a later time than s. The metric tensors in $\mathbb{S}_x^{g(x)}$ at the points $y_{g(x)} = cs, \mathbf{y}$ at time s are given by parallel transport of $[\beta_{cs}(y)]_{g(y)} = [\beta_{cs}(cs)]_{g(y)}$ to x. The result is

$$[\beta_{ct}(cs)]_{g(x)} = [e^{-\alpha(ct)+\alpha(cs)}\beta_{cs}(cs)]_{g(x)} = [e^{-\alpha(ct)+\alpha(cs)}\eta]_{g(x)}. \tag{6.189}$$

The equation shows that if $\alpha(cs) > \alpha(ct)$, then $\beta_{ct}(cs) > \beta_{cs}(cs)$. If $\alpha(cs) < \alpha(ct)$, then $\beta_{ct}(cs) < \beta_{cs}(cs)$.

Let $d\vec{y}$ be a differential space vector at location $y = cs, \mathbf{y}$ in M. Lift of this vector to an image location, $y_{g(y)}$, in the local space, $\mathbb{S}_y^{g(y)}$, gives the vector $d\vec{y}_{g(y)}$ in the local space. In a similar way, lift of the space vector $d\vec{x}$ at position, $x = ct, \mathbf{x}$ in M gives the vector, $d\vec{x}_{g(x)}$, in the local space $\mathbb{S}_x^{g(x)}$.

The lengths of the differential vectors in their respective spaces are given by

$$|d\vec{y}|_{g(y)} = [(\beta_{cs,\mu,\nu}(y)dy^\mu dy^\nu)^{1/2}]_{g(y)} = [(\eta_{\mu,\nu}dy^\mu dy^\nu)^{1/2}]_{g(y)} \tag{6.190}$$

and

$$|d\vec{x}|_{g(x)} = [(\beta_{ct,\mu,\nu}(x)dx^\mu dx^\nu)^{1/2}]_{g(x)} = [(\eta_{\mu,\nu}dx^\mu dx^\nu)^{1/2}]_{g(x)}. \tag{6.191}$$

These lengths are numbers in $\bar{R}_y^{g(y)}$ and $\bar{R}_x^{g(x)}$. These equations have used the fact that $\beta_{cs,\mu,\nu}(y)$ and $\beta_{ct,\mu,\nu}(x)$ are both equal to $\eta_{\mu,\nu}$.

Parallel transport of the differential length $|d\vec{y}|_{g(y)}$ to location $y_{g(x)}$ in $\mathbb{S}_x^{g(x)}$ gives the result

$$
\begin{aligned}
C_g(x,y)|d\vec{y}|_{g(y)} &= [e^{-\alpha(ct)+\alpha(cs)}|d\vec{y}|]_{g(x)} \\
&= [\beta_{ct,\mu,\nu}(y)(dy^\mu dy^\nu)^{1/2}]_{g(x)}.
\end{aligned}
\tag{6.192}
$$

Here

$$
\beta_{ct,\mu,\nu}(y) = e^{-2\alpha(ct)+2\alpha(cs)}\eta_{\mu,\nu}.
\tag{6.193}
$$

The properties of the connection show that $|d\vec{y}|_{g(y)}$ is the same length number in $\bar{R}_y^{g(y)}$ as $[\beta_{ct,\mu,\nu}(y)(dy^\mu dy^\nu)^{1/2}]_{g(x)}$ is in $\bar{R}_x^{g(x)}$. Their length values are different.

If α has the time dependence shown in Figure 6.20, then $\alpha(cs) > \alpha(ct)$. Equation (6.192) shows that the local length and distance values increase as the distance, $c(t-s)$ between points, y and x in M, increases. The value of the length, $|d\vec{y}|_{g(x)}$, at location $y_{g(x)}$ in $\mathbb{S}_x^{g(x)}$ is larger than the value of the length, $|d\vec{y}|_{g(y)}$, at the location $y_{g(y)}$ in $\mathbb{S}_y^{g(y)}$.

This relation and the magnitude of the difference between the lengths in the local spaces at y and x are independent of the space location, $y = \mathbf{y}$, at the space–time location, $cs, \mathbf{y}$. One concludes from this that the local space–time $\mathbb{S}_x^{g(x)}$ is an expansion of $\mathbb{S}_y^{g(y)}$. If $\alpha(cs) > \alpha(ct)$ for all times s from $s = 0$, the big bang time, to $s = t$, Figure 6.20, then the space–times are continually expanding as time s increases. Conversely, if $\alpha(cs) < \alpha(ct)$, then the local space–time at x is a contraction of the space–time at y.

The rate of space–time expansion or contraction as the time s increases is given by

$$
\begin{aligned}
\frac{d}{ds}|d\vec{y}|_{g(y)} &= \frac{d}{ds}[e^{\alpha(cs)-\alpha(x)}|d\vec{y}|]_{g(x)} \\
&= [-c\vec{A}(cs)]_{g(x)}[e^{\alpha(cs)-\alpha(x)}|d\vec{y}|]_{g(x)} = [-c\vec{A}(cs)]_{g(x)}|d\vec{y}|_{g(x)} \\
&= [-c\vec{A}(cs)|d\vec{y}|]_{g(x)}.
\end{aligned}
\tag{6.194}
$$

This equation shows that if $\vec{A}(cs)$ is positive, or negative, then the space–time at x is an expansion, or contraction, of the space–time at y. If $\vec{A}(cs)$ is positive for $s < t$ and Eq. (6.181) for small red shifts is valid, then the rate of expansion of the local space–times as s increases can be equal to the Hubble parameter. For the time dependence of α shown in Figure 6.20, the rate of expansion of $\mathbb{S}_y^{g(y)}$ as s increases for times later than 4 billion years ago accelerates from the constant rate at earlier times. This acceleration of expansion can be ascribed to the effect of dark energy.

With the functional decrease of α as time increases, as exemplified in Figure 6.20, the space components of the space–times at all space locations at a light cone distance $c(t-s)$ from x expand with time at the same rate, $-\vec{A}(cs)$. If $-A(cs)$ equals the Hubble parameter, as has been suggested, then the expansion rate of space is very small. From Eq. (6.184), one sees that the fractional rate of expansion is about 7×10^{-11} per year. This is extremely small. It is so small that there is no hope of observing this effect, as a red shift, for light emitted from local sources.

The upturn of the curve in Figure 6.20 for small s shows a decreasing rate of expansion of the spaces, $\mathbb{S}_y^{g(y)}$, as s increases. This is consistent with the assumption that $\alpha(cs) \to \infty$ and $\vec{A}(cs) \to \infty$ as s approaches 0. From the viewpoint of an observer at x, the space components of the space times, $\mathbb{S}_y^{g(y)}$, get smaller as s decreases. As s approaches, 0, the big bang time, the space components of $\mathbb{S}_y^{g(y)}$ at each space location, $\mathbf{y}_{g(x)}$, in $\mathbb{S}_x^{g(x)}$ approach the limit of having zero spatial extent. The spaces shrink to a point at $s = 0$. The energy of matter in the space–time approaches infinity as $s \to 0$. This follows from the shrinking of the wavelengths of radiation and matter to 0 as $s \to 0$.

If the time dependence of α is such that it can describe the big bang at $s = 0$, and inflation occurs, then α might be expected to show the expansion due to inflation. This expansion rate is very rapid. It is described to begin at a time 10^{-33} s after the big bang and to end at time 10^{-32} s. During this time, lengths expand by a factor of 10^{26} [76]. This corresponds to an increase in the exponential scale factor, $e^{\alpha(cs)}$, of the metric tensor β_{cs} at $s = 10^{-33}$ s to the tensor, β_{cs} at $s = 10^{-32}$ s by a factor of 10^{26}.

Another feature of the time dependence of $\alpha(cs)$ is that it should show the rate of expansion of space from the time of recombination to the present. This corresponds to requiring that the rate of change, $= \vec{A}(cs)$, of the radiation from a red shift of about 1,100 at the recombination time of about 38,000 years [78] to the present time where the red shift is 0.

The zero red shift at the present time, t, is a statement about the properties of the metric tensor in the local space–time, $\mathbb{S}_z^{g(z)}$, where $z = ct, \mathbf{z}$. It is equivalent to the statement that $\beta_{ct}(ct) = \eta$.

If $\alpha(cs) \to \infty$, as suggested in Figure 6.20, then the metric tensor for the local space at time t approaches infinity. That is, $\beta_{ct}(cs) \to \infty$ as $s \to 0$. This and the fact that the wavelength of radiation in the cosmic microwave background is finite show that the spatial extent of the space–times in the past light cone of z approaches 0 as s approaches 0, the time of the big bang. Since the wavelengths of radiation and material matter shrink to 0, the energy of the radiation and matter become infinite. This shows that the big bang is an infinite energy source.

So far in this chapter, it has been shown what the value field α can do. It can account for the expansion rate of space that corresponds to that of the Hubble parameter. It can account for the acceleration of space expansion due to dark energy. It may also account for the big bang at $s = 0$. This does not mean it must do these things. The discovery of additional requirements that α must have awaits future work. One may hope that these requirements show that the value field must have the properties outlined above.

Chapter 7

The Theory Experiment Connection

7.1 Theory predictions and experiment outcomes

Theories are mathematical models of physical systems and processes. Systems, such as protons, molecules, and others, and processes, such as dynamics and scattering events and large-scale cosmological properties, are represented by mathematical elements in the models. These mathematical models are usually represented with mathematical elements that belong to global mathematical structures. Global mathematical structures exist independent of space or time. They are not associated with any location or region of space, time, or space–time. This is the mathematics that is in general use in physics and mathematics.

In the framework of local mathematics and the presence of a value field, theories, as mathematical models of physical systems, are contained in the local mathematics and local space–times at points in M. For nonrelativistic physics, the local and global space–times become spaces.

The validity of any physical theory rests on its experimental support. This is true for theories based on either global or local mathematics. The theory must be able to explain results of completed experiments as well as predict outcomes of experiments that have not yet been done. The predictions of a successful theory must turn out to be supported by experiments whenever they are done. The strength of a theory is based on the number and significance of experiment outcomes it can describe and predict.

As has been seen from the previous chapters, local mathematics and the presence of the value field affect theoretical descriptions of many physical and geometric quantities. This includes quantities that are represented by derivatives or integrals over space, time, or space–time. This applies to small systems, such as those described in quantum mechanics, and to large cosmological systems. The latter is exemplified by the use of the α field to describe the cosmological red shifts due to the Hubble expansion and possibly dark energy.

As is well-known, theories stand or fall on the basis of experimental support or refutation. A main purpose of this short chapter is to investigate how the presence of local mathematics and the value field affect the experimental verification or refutation of theories.

Theory predictions are implemented by computations. These and experiments are physical processes. Each theory prediction is associated with a specific program in a computer. In essence, the program gives the protocol or instruction set for the computer. The computation is a physical process that halts. The output is registered in a designated system that may or may not be part of the computer.

For experiments, the process can be described as a sequence of steps carried out in the experiment. The specifics of the experimental process are determined by a set of instructions or protocol. As is the case for computations, the protocol for each experiment is such that the experimental process halts. The outcome of the experiment is registered in some specified system.

Computer outputs and experimental outcomes are physical systems in certain states. These states are usually interpreted as magnitudes of different physical quantities. These magnitudes are represented by numbers. The interpretation of output and outcome states as numbers is provided by the physical form of the output and outcome states and knowledge of what computers do and of what an experiment is supposed to accomplish.

There are many physical forms for these output systems. They can appear as strings of digits written on tape or in some optical or acoustic form. The form used must be such that each output state must have an interpretation as a number of some type.

A basic character of all numerical experiment or measurement outcomes and computer outputs is that they represent natural numbers, integers, or rational numbers. A rational number output or outcome does not represent a specific real or complex number.

This is a consequence of the fact that the physical outcomes and outputs contain a finite amount of information. Symbol string representations of outcomes and outputs contain at most a finite number of digits. Furthermore, the number of digits is not large.

In the arena of local mathematics and the presence of a value field, knowledge of the computer output and experimental outcome numbers is a necessary but not sufficient condition for determining if theory agrees or disagrees with experiment. The reason for lack of sufficiency is based on the observation that comparison of theory with experiment to determine agreement of disagreement has meaning or value to an observer. This comparison is done by comparing the meaning or value of the computation output number with that of the experimental outcome number.

It is worthwhile to give some details for the steps needed to determine if theory is supported or refuted by experiment. Let a be the number associated with the theory computation output state and x be the location of the computation. This number can be represented by a number with value, b, as in

$$a = b_{g(x)}. \tag{7.1}$$

Here $g(x) = e^{\alpha(x)}$ is the scale factor for the local number structure at x that contains the computation number.

Let d be the number associated with an experiment outcome state and y be the location of the experiment outcome state. The experimental outcome number, d, can be represented by the number $e_{g(y)}$ as in

$$d = e_{g(y)}. \tag{7.2}$$

Here y is the location of the experiment and e is the number value of the experimental outcome. The scale factor for the number structure at y is given by $g(y) = e^{\alpha(y)}$.

Figure 7.1 gives a schematic layout for a computer, with its output and an experiment with its outcome. This is the basic setup considered for theory experiment tests for agreement or disagreement. Computer output and experiment outcome states are represented in the figure as symbol strings. These strings are interpreted as numbers. The values of these numbers are determined by the local number structures containing the numbers and by the value factors, $g(x)$ and $g(y)$, for the structures at x and y.

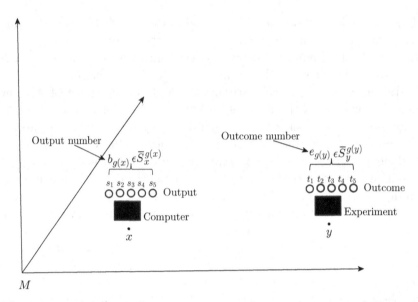

Figure 7.1. Schematic for representations of computer outputs and experimental outcomes for comparison in a test of agreement between theory and experiment. The black boxes represent a computer at location x and experimental apparatus at location y. The computation output and experimental outcome are shown as symbol strings $s_1 \ldots, s_5$ and $t_1, \ldots, t_5$. These strings are interpreted to be the numbers $b_{g(x)}$ and $e_{g(y)}$. These represent numbers in the number structures, $\bar{S}_x^{g(x)}$ and $\bar{S}_y^{g(y)}$, with respective number values, b and e. The structure label, S, denotes the number type (integer, rational, and real) appropriate for the theory experiment test under consideration.

In this chapter, local number structures are denoted by $\bar{S}_x^{g(x)}$ and $\bar{S}_y^{g(y)}$. They refer to the type of number (natural, integer, rational, and real) represented by the computer output and experiment outcome.

Determination of agreement or disagreement between theory and experiment requires comparison of the values of the numbers a and d. However, the "no information at a distance" principle [16,17] forbids direct comparison of the values of these numbers as elements of structures at different locations.

This problem can be fixed by physical transport of the information contained in the output and outcome states to a reference location, z, for comparison. This can be implemented by the use of light as a transport medium or copying the numbers onto paper and bringing the paper copies to z. These are only two of the many methods available. Since the physical systems for the theory computation output and experiment outcome are at the same location, z, their value in the number structure at z can be compared.

The values or meanings of these numbers change as they are moved along a path from x and y to z. The change can be represented by use of the value functions introduced in Section 1.3 on number scaling. The change can also be described by parallel transport of the numbers from x and y to z.

The values of the computer output and experimental outcome numbers, a and d at their locations of production, are given by $v_{g(x)}(a)$ and $v_{g(y)}(d)$. The values of these numbers after physical translation to z are $v_{g(z)}(a)$ and $v_{g(z)}(d)$. From Eqs. (7.1) and (7.2), one sees that these values can be written in an equivalent form as $v_{g(z)}(b_{g(x)})$ and $v_{g(z)}(e_{g(x)})$. The value changes resulting from physical transport are represented mathematically by parallel transport of the numbers to z. This is given by

$$[v_{g(z)}(a)]_{g(z)} = [v_{g(z)}(b_{g(x)})]_{g(z)} = C_g(z,x)b_{g(x)} = [e^{-\alpha(z)+\alpha(x)}b]_{g(z)}$$
$$(7.3)$$

and

$$[v_{g(z)}(d)]_{g(y)} = [v_{g(z)}(e_{g(y)})]_{g(z)} = C_g(z,y)e_{g(y)} = [e^{-\alpha(z)+\alpha(y)}e]_{g(z)}.$$
$$(7.4)$$

Figure 7.2 is a schematic representation of the relations between physical and mathematical transports of the theory and experimental outcomes. The figure shows that the two methods of transport, physical transport of the numbers followed by valuation, give the same numbers with the same values as those obtained by valuation followed by parallel transport. The fact that both methods give the same result is a consequence of the number preserving property of physical transport and of parallel transport.

The figure shows the setup for two-dimensional Euclidean space. In space–time, the location z would have to be timelike relative to x and y.

Determination of agreement or disagreement between theory and experiment requires comparison of the values of the numbers

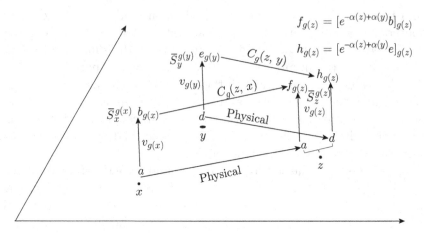

Figure 7.2. Physical transport of numbers followed by valuation gives the same numbers with the same values as those obtained by valuation followed by parallel transport. The arrows labeled "physical" show physical transport of the numbers, a and d, from sites, x and y, to z. The vertical arrows show the local number structures containing the numbers and the effect of the value function to give representations of the numbers as numbers with specified values in the structures. Parallel transport of these number representations is shown by the upper arrows with connection labels, $C_g(z,x)$ and $C_g(z,y)$.

a and d at z. If $e^{-\alpha(z)+\alpha(x)}b$ differs from $e^{-\alpha(z)+\alpha(y)}e$ by less than statistical uncertainties, then experiment supports theory. If the difference is outside statistical uncertainties, then experiment refutes theory.

Equations (7.3) and (7.4) show that the values of the computation output and experimental outcome numbers can depend on the locations of the computation and experiment. This is a result of the presence of $\alpha(y)$ and $\alpha(x)$ in the exponentials. If α varies over the range of possible locations of the computations and experiment, one can have the situation that agreement or refutation of theory by experiment can depend on the locations of the computation and experiment. At some locations, experiment supports theory. At others, it refutes theory.

This is not a problem if the variation in the values of α over the possible locations of the computation and experiment are so small that the location dependence of the number values is less than the uncertainties associated with the computation and experiment. If α is a constant over the possible locations of the computation and experiment, then this problem disappears. In this case, there is no location dependence. This is a consequence of the fact that parallel transports, implemented by connections, preserve both number and number value.

7.1.1 *Statistical theories*

The description given above of theory experiment comparison for a single computer output with a single experiment outcome can be extended to statistical theories. Quantum mechanics provides many examples in that a computation of an observable expectation value is compared with the mean of repeated measurements of the observable on particles prepared in the same state but at different times and/or locations. The measurement of the momentum of a quantum particle is an example.

As noted in Chapter 5, in the framework of local mathematical structures and with the α field present, a particle with wave function $\psi(u)_{g(u)}$, in the complex number structure, $\bar{C}_u^{g(u)}$, at each point u becomes the wave function

$$\phi(u)_{g(z)} = [e^{-\alpha(z)+\alpha(u)}\psi(u)]_{g(z)} \qquad (7.5)$$

in the complex number structure, $\bar{C}_z^{g(z)}$, at reference point z. The momentum wave function at z is given by

$$\theta(p)_{g(z)} = \left[\frac{1}{(2\pi)^{3/2}} \int e^{ipu}\phi(u)du\right]_{g(z)}. \qquad (7.6)$$

The momentum expectation is given by

$$[\langle\theta|\tilde{P}|\theta\rangle]_{g(z)} = \left[\int |\theta(p)|^2 pdp\right]_{g(z)}. \qquad (7.7)$$

The quantity within the square brackets is the momentum expectation value.

Measurement of the momentum expectation consists of repeated preparations of particles in the state, θ, at the same or different locations followed by measurements of the momentum of each particle. The mean of the experimental outcomes is to be compared with a computation output of the calculation of the expectation value.

Experiment supports theory if the difference between the values of the computation output and experimental mean is less than the combined uncertainty values from the theory computation and experiment procedures.

The effects of α can be seen from a specific example. Let x be the location of a computation of $[\langle\theta|\tilde{P}|\theta\rangle]_{g(x)}$ and $y_1,\ldots,y_n$ be the locations of n repeated measurements of the momentum. Let $r_{g(x)}$ be the computation output number and $[p_1]_{g(y_1)}\cdots[p_n]_{g(y_n)}$ be the n experiment outcome numbers. For each $j = 1,\ldots,n$, $[p_j]_{g(y_j)}$ is a rational number approximation to a real number in $\bar{R}_{y_j}^{g(y_j)}$ and $r_{g(x)}$ is a rational number approximation to a real number in $\bar{R}_x^{g(x)}$.

Determination of agreement between the computer output and the mean of the experiment outcomes requires transport of the computer and experiment outcome numbers to a common location, z, for comparison. Parallel transport of the computer output to z gives the number, $a_{g(z)} = [e^{-\alpha(z)+\alpha(x)}r]_{g(z)}$. Parallel transport of the n experiment outcome momentum numbers to z gives the n numbers

$$[q_j(y_j)]_{g(z)} = [e^{-\alpha(z)+\alpha(y_j)}p_j]_{g(z)},$$

where $j = 1,\ldots,n$. The mean of the n experimental outcomes is given by

$$M(p,n)_{g(z)} = \left[\frac{1}{n}\sum_{j=1}^{n}q_j(y_j)\right]_{g(z)} = \left[\frac{1}{n}\sum_{j=1}^{n}e^{-\alpha(z)+\alpha(y_j)}p_j\right]_{g(z)}. \tag{7.8}$$

The standard deviation associated with the mean is given by

$$\sigma(p,n)_{g(z)} = \left[\sqrt{\frac{\sum_j(q_j(y_j) - M(p,n))^2}{n}}\right]_{g(z)}. \tag{7.9}$$

Let $\Delta_{g(x)}^{\text{th}}$ be the uncertainty of the computation of the momentum expectation number. This includes the uncertainty that the program for the computation accurately represents the theoretical prediction and uncertainties due to errors in the steps of the physical computation and to roundoff errors. Let $\delta^{\text{th}}(x)_{g(z)} = [e^{-\alpha(z)+\alpha(x)}\Delta^{\text{th}}]_{g(z)}$ represent the transported computation uncertainty to z.

Determination of experimental support or refutation of the theory prediction is done here for number ranges instead of number value ranges. As noted before, it is immaterial whether comparison between numbers or number values is done, provided the comparison is made within a single number structure. The determination is obtained by looking for overlap between the number intervals

$$a_{g(z)} \pm [\delta^{th}(x)]_{g(z)} = [a \pm \delta^{th}]_{g(z)} \tag{7.10}$$

and

$$M(s,n)_{g(z)} \pm \sigma(s,n)_{g(z)} = [M(s,n) \pm \sigma(s,n)]_{g(z)}. \tag{7.11}$$

If there is no overlap, one concludes that experiment does not support theory. If there is overlap, then the experiment does support the theory. These conclusions are valid to an uncertainty of one standard deviation.

The variation of the α field has the consequence that agreement or disagreement between theory and experiment can depend on the locations of the n experimental repetitions and on the location of the computer output. This is especially the case if the α field varies a lot between the locations of the experiments and the computation. As will be seen in Section 7.3, the variations are too small locally to have this effect.

7.2 Experiments, measurements, and computations

At this point, it is necessary to face up to the fact that, so far, there is no direct experimental evidence for the presence of a valuation field. It has not been used in theoretical expressions for physical quantities or in the comparisons between theory predictions and experimental results. This does not mean that a space or time varying α or the gradient field, $\vec{A}$, does not exist and affect physical quantities. Examples of the effect are given in Section 6.6.1 on α and the cosmological red shift.

Reconciliation of these two aspects of the effect of the α field in physics is a consequence of the distinction between experiments and measurements as the terms are used here. It is also a consequence of some properties that all experiments, measurements, and theoretical computations have in common.

7.2.1 *Experiments differ from measurements*

Experiments can be distinguished from measurements in that experiments include state preparation procedures. A system whose properties are to be measured is prepared in some state. Then a measurement is done on the state of the system to determine whether the system has some property or not. Or the measurement may determine the magnitude of some physical quantity that the system has. For statistical theories such as quantum mechanics, this process of preparation followed by measurement is repeated many times.

There are many examples of experiments. Preparations of quantum systems in some states followed by measurements of properties such as excitation energies, momentum distributions, and spin are examples. Use of the Large Hadron Collider to prepare very high energy proton beams followed by measurements of the proton proton collision products is another example.

As is noted above, all experiments include a measurement of properties of a system prepared in some state. Measurements are more general than experiments in that measurements can be made on properties of systems where there is no preparation step involved. Properties of the system are determined by examining properties of the radiation of various types emitted by the system or systems. The increased generality of measurements over experiments is that they include determination of properties of systems that are very large or very far away. Examples include measurements of properties of large systems, such as the Earth, other planets, and the Sun. Larger cosmological systems such as galaxies and black holes are other examples of these systems.

Properties of these systems are determined by properties of radiation from the systems as they are. For the most part, the radiation is electromagnetic. This includes light, radio waves, X-rays, and gamma rays. Use of other radiation types, such as cosmic rays and neutrinos, to determine properties of systems is coming into use [79]. The point here is that the properties such as size, distance, and relative velocities are inferred from properties of the incoming radiation. Cosmological red shifts are an example.

7.2.2 Experiments, measurements, and computations are local

A basic aspect of all experiments, measurements, and computations is that they are implemented locally. The preparation of a system for measurement is done locally with physical equipment assembled for the purpose. Measurements are carried out by physical equipment or apparatuses collected locally for the measurement. Similarly, computations are done by local physical computers.

The local restriction is a consequence of the fact that experiments, measurements, and computations are carried out by us as observers. We are the ones that provide equipment for implementing experiments, measurements, and computations. We are the ones that justify the observation that the assembled equipment and computers do what they are intended to do. We are the ones who test for agreement between theory and experiment.

The fact that we as observers carry out experiments, measurements, and computations has the consequence that these operations are limited to regions of space–time that are occupied by us. It also includes regions of space–time that are occupiable by us, now or in the future. These regions include much of the space–time region occupied by the solar system. There is no reason why we cannot, in principle, set up space stations almost anywhere in the solar system and carry out experiments to test theory. The international space station is a step in this direction. The Voyager space ships show that the region can be extended out beyond the solar system. However, the region is limited to be a small fraction of the size of the universe.

7.3 Experimental limits on the α field

Experimental limits on the α field are based on the fact that, so far, no experiment has given an outcome that has been interpreted to show the presence of the α field or its gradient. This lack of experimental evidence places limitations on the properties of α and its gradient vector field, $\vec{A}$. These limitations are local in that all state preparation procedures, measurement procedures, and computations are necessarily implemented by us as observers. All these procedures are limited to the regions of space–time that are occupied or occupiable by us.

The same locality condition holds for theoretical predictions. These are implemented by computations. Since computations are also implemented by us, they are restricted to regions of space–time that are occupied or occupiable by us as observers. These conditions also apply to robots that are or may be developed to carry out experiments and computations.

The emphasis here is that, to date, no *experiment* has given results whose interpretation includes the value field or its gradient. The reason for this is that the preparation step, as an essential part of an experiment, is necessarily implemented locally by us. The source of the emitted radiation, particulate to electromagnetic, is local. The radiation does not have to travel far from the source to the detector. The source to detector distance is limited by the fact that both the source and detector must be in the region of space–time occupiable by us.

It is also the case that, to date, no measurement has been done whose results have been interpreted to include the value field or its gradient for explanation. However, measurements do not include a preparation step. Measurements determine properties of systems or radiation incoming to the measurement apparatus or detectors.[1]

A consequence of this distinction between experiments and measurements is that, for measurements, there is no limitation on the source to detector distance. The source can be anywhere in the universe. The detector must be local as it includes locally constructed hardware. The properties of the source are determined by analyzing the properties of the incoming radiation emitted by the source.

The fact that, for measurements, the source to detector distance can be arbitrarily large has the consequence that the value field and its gradient can be used to explain the results of measurements on radiation from cosmological sources. This was done in the last chapter where variations in the α field were used to account for the red shift due to Hubble expansion and the acceleration effect of dark energy.

---

[1]Measurements can include equipment to shape or change the incoming systems. However, preparation of the source of the emitted systems is not included.

The limitation on the α field to be unobservable in experiments anywhere in the region of the universe occupiable by us is too restrictive. The reason is that, so far, experiments have been carried out by us near, or on, the surface of the earth. This is a small part of the occupiable region. One does not know if experiments set up in a laboratory near or on Pluto would give results whose interpretation requires the use of the value field, α.

Even so, it seems prudent to assume the effects of the α field would be unobservable in experiments carried out anywhere in the region of the universe occupiable by us. If some future experiments carried out by us give results that use the value field in their interpretation, then the size of the restricted region would have to change.

The observation that variations in the α field are undetectable in any experiment done by us so far can be given a more precise description. Let x and y be two locations in the space–time region occupiable by us. As has been seen before, the connection that parallel transports a meaningful physical quantity, q, from y to x is given by

$$q_{g(y)} \to C_g(x, y)q_{g(y)} = [e^{-\alpha(x)+\alpha(y)}]_{g(x)}q_{g(x)}. \qquad (7.12)$$

The undetectable condition is given by the requirement that the difference, $\alpha(y) - \alpha(x)$, is smaller than the statistical uncertainty associated with the outcome of any experiment done so far that measures the property q. This restriction on $\alpha(y) - \alpha(x)$ must hold for all properties q and for all point pairs y, x in the region of space–time occupiable by us. The additional requirement that x is not spacelike relative to y is also needed.

This condition can also be expressed by the number value equation:

$$e^{-\alpha(x)+\alpha(y)} \approx 1 + \alpha(y) - \alpha(x). \qquad (7.13)$$

This equation shows that if $\alpha(y) = \alpha(x)$, then the connection is the identity.

The case where $\alpha(y) = \alpha(x)$, and the connection is the identity, is independent of the use of local mathematics in physics and elsewhere. In this case, theoretical predictions and experiment outcomes and their interpretation are the same as those in the usual physics based on global mathematics.

The description of the limitations on the variation of α for regions of space and time occupiable by us leaves out an important observation. This is the possibility that observers, or intelligent beings, capable of doing experiments and testing theory, may, in principle at least, exist almost everywhere in the cosmological universe.

However, determination of whether their physics agrees or disagrees with ours is possible only if they are close enough to us so that effective two way communication is possible. This communication is necessary to determine if the physics created by the extraterrestrial beings agrees with ours. This includes determination whether or not existence of the α field variations is needed by these intelligent beings to interpret their results.

So far, it is not known if experiments carried out on planets around stars other than the Sun give results that restrict the variation of α. For this reason, it seems prudent to assume that the restricted region must be sufficiently large to include all regions occupiable by us and by other intelligent beings for which two-way communication is possible.

7.3.1   *The size of the region of restrictions on α*

There are two conditions that the size of the region of α restriction must satisfy. One is that the far away beings must be sufficiently close that any signal they send to us must be distinguishable from background noise. This gets increasingly difficult with increasing distance. This problem is exemplified by the observation that, so far, the search of extraterrestrial beings (SETI) [80] has turned up empty.

The other limitation on the size of the restricted region is a consequence of the finite speed of light. The size of the region is limited by the fact that the time required between two way conversations between us and intelligent beings, if any, in the region is not too large. One concludes from this that the size of the restricted region must be very small compared to the 14 billion light year size of the universe.

One estimate [81] of the size of this region is that it extends about 1,000 light years from us. For our purposes, this is an overly generous restriction because it is an estimate of the region size in which we can determine whether or not intelligent beings exist.

Outside the restricted region, there are no limitations on the variation of the valuation field. It can vary very rapidly or not at all. The field could be such that an experiment done by an observer at a million light years from earth supports a very different theory in physics than our theory. Such a possibility may make an interesting speculation, but it is outside the domain of physical theory support or refutation by us or civilizations on planets close enough so that two-way communication is possible. Statements about what such a distant observer might find by experiments are neither true nor false. They are meaningless because communication with these observers is impossible.[2]

These conclusions do not affect the fact that astrophysics and cosmological physics are well-established fields of physics with much observational support. This includes determining properties of red shifted systems, observation of gravitational lensing, pulsars, and many other types of systems.

It is important to note that the observational support for such theories is local. The measurements and observations of properties of these distant objects are carried out by us as observers in our local restricted region. Nevertheless, these local measurements are used to support or refute theoretical models or descriptions of distant systems. The success of a theoretical model is based on the number of predictions that are supported by local measurements.

7.4   The existence of α as a physical field

Nothing said so far precludes the existence of α as a physical field. The restrictions imposed so far require that variations of α in the local regions of space and time be sufficiently small to be unobservable. Outside this region, there are no restrictions on the variation of α or on the strength of the gradient vector field.

There are many fields proposed in the physics literature that have no effect locally but have an effect globally. As is the case with the

---

[2]This restriction would have to be revised if, in the future, a method was found to transmit information at speeds greater than that of light. In addition, one would have to be able to distinguish the signals from such distant beings from noise.

α field, these fields, such as the inflaton [82], quintessence [83,84], and other proposed fields [85], are scalar.

None of these fields affect the outputs of computations or outcomes of experiments or measurements. The physical processes of computations do not have to be designed to account for the local presence of these fields. Experimental equipment does not have to be designed to be protected or shielded from local effects of these fields. There either is no local effect or the effect is way too small to be observable.

The red shift is an example of this. The Hubble expansion affects all systems whether they are nearby or far away. For example, a source of light one meter away from a detector is red shifted. But the effect is far too small to be detected. The amount can be determined from the Hubble parameter of about 70 km/sec/megaparsec [75]. A change of units gives the Hubble parameter as

$$70 \, \text{km/s/megaparsec} \simeq 2.7 \times 10^{-18} \, \text{m/s/m}.$$

This is far too small to have any effect on local experiments or measurements of local quantities.

Fields such as the inflaton, quintessence, and other scalar fields are proposed as components of theoretical models of global or large scale properties of the universe and its evolution. As such they are used to interpret the results of local measurements. The same properties hold for dark matter [86,87] and dark energy [88–90]. These quantities as fields have no effect on the designs or setups of equipment used for local computations or experiments.

To be specific, suppose a computation or experiment is carried out at space–time location u. The experiment is repeated at another location v. The fact that the dark matter field may be different at v than at u or that space–time has expanded at v relative to space–time at u can be ignored. If they exist at all, these effects are far too small to have any effect on computations or experiments.

The possibility that the α field may be part of the theoretical model describing the evolution of the universe is described in Section 6.6.1. There it is shown that the α field can describe the rate of red shift increase with distance that corresponds to the Hubble expansion. It was also seen that the field can also describe the acceleration of the rate of red shift with distance that corresponds to the effect of dark energy. The more speculative possibility that the α field can be used to describe the big bang was also briefly noted.

The α field is a scalar field presumably of spin 0. This suggests the possibility that α is related to or is the Higgs field. In this case, the observations in Section 6.6.1 that variations in α can show the effect of the Hubble expansion and dark energy are of interest. They suggest that the field $\vec{A}$ may be the source of both the Higgs potential and the red shifts from the Hubble expansions and dark energy. This suggestion adds to other work on possible relations between the Higgs field and dark energy [91]. Whether the suggestion described above for $\vec{A}$ is valid or not is not known at present. More work needs to be done on this topic.

Work also needs to be done to determine if the α field is related to other scalar fields, such as quintessence or the inflaton. At present, any relation is pure speculation.

Chapter 8

The Basic Role of the α Field

8.1 Introduction

From the work presented so far, one would be forgiven for thinking that the α field and its gradient play at most a minor role in local physics and mathematics. Variations in the α field over the local cosmological region containing us and possibly other intelligent beings for which two-way communication is possible must be too small to affect either theory descriptions or predictions in physics and their support or refutation by experiments and measurements. So far, the effects of the gradient, $\vec{A}$, of α on physics have been limited to giving a possible description of the red shift and dark energy. The possible relation to the Higgs field was also noted.

Nevertheless, the α field is essential to physics and mathematics. It may even be essential to the relation between conscious observers and the physical universe. Support for this is discussed in the following subsection. It is based on the observation that computer outputs as theory predictions and experiment outcomes are physical systems in physical states. As such, they are numbers. However, they have no intrinsic value or meaning. Any value is possible. This problem is remedied by the presence of the value field, α.

8.1.1 *Numbers have no intrinsic meaning*

One begins with the observation that comparison of theory predictions with experimental results requires that theory predictions are implemented in a way that is compatible with outcomes of experiments. A much used method of achieving this is to represent theory predictions as numbers generated by computers. Similarly, experiments are often arranged so that their outcomes are also numbers.

At the most basic level, computer outputs and experimental outcomes are physical systems in different physical states. They have no intrinsic meaning or value. However, the form of these physical states makes them recognizable as numbers. For example, they can be strings of symbols recognizable as digits. They can be encoded into light or sound waves with appropriate frequency and amplitude variations.

The fact that the form of the output and outcome states makes them recognizable as numbers is a necessary condition for the output states to have numerical value. However, as has been repeatedly emphasized in this book, it is not sufficient as it does not imply that the physical systems in these states, as numbers, have meaning or value. Number and number value are distinct concepts. For example, 3.42 is a possible output or outcome number. But, as has been emphasized before, it has no particular value. Any value is possible.

The reason for this lack of meaning is that there are an infinite number of rational or real number structures collocated with a computer output or experimental outcome. The value or meaning of the number depends on the value factor for the structure providing the environment of the number.

The essential role of the value field α is to provide value factors for the structures containing output or outcome numbers. The value or meaning of the number is determined by its properties in the collocated number structure. Without the α field present, the values or meanings of the computer output and experimental outcome numbers are unknown.

8.1.2 α and extended local mathematics

The importance of α in providing meaning to numbers and other types of mathematical systems, containing numbers as part of their description, can be illustrated by considering an extension of local mathematics. In this extension, the set of mathematical systems at location, x, in M is expanded from $\cup_S \bar{S}_x$ to

$$G_x = (\cup_r(\cup_S \bar{S}_x^r)). \tag{8.1}$$

It is clear from this extended representation of local mathematical systems that a number of type S at x has no meaning or value. Any value is possible. The value depends on the value or meaning factor of the structure containing the number. Scalar properties of vectors have no meaning or value. Any values are possible.

The problem this causes for assigning meaning or value to computer output numbers and experimental outcome numbers is shown in Figure 8.1. This figure shows the situation in the absence of a value field for computer outputs and experimental outcomes as numbers. The numerical output of a computer at x and an experimental outcome at y are shown as a and d. The infinite number of number structures available at locations, x and y, are shown as columns of number structures. Two of the infinite number of possible choices of number structures as environments for the output and outcome numbers are shown.

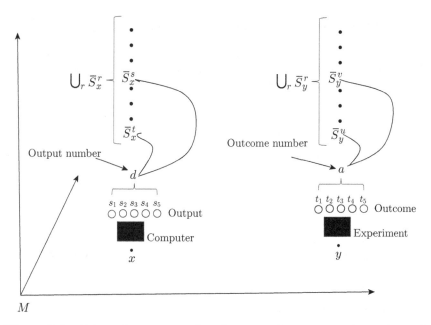

Figure 8.1. Schematic representation of computation output and experimental outcome states at locations x and y. The numbers represented by these states are shown as a and d. The vertical dots with number structures $\bar{S}_x$ and $\bar{S}_y$ denote either local rational or real number structures at x and y. Two of the infinite number of number structures to which the numbers a and d can belong are indicated by the curved lines. The values or meanings of the numbers, a and d, depend on the value factor of the structure containing them.

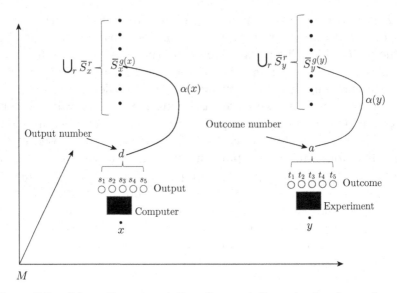

Figure 8.2. Schematic representation of computation output and experimental outcome states at locations x and y. The numbers represented by these states are shown as a and d. The infinite collections of number structures of type S at x and y are indicated by the vertical dots. The unique structures of type S, chosen by the values of the α field at x and y, are shown by the curved lines labeled with α.

The figure shows if d is a number in $\bar{S}^t$ its value is c. If d is a number in $\bar{S}^s$, then its value is k. The relation between the two number values is given by $k = (t/s)c$. Similarly, if $a = b_u = j_v$, then $j = (u/v)b$.

Figure 8.2 illustrates the situation in the presence of the value field for numbers at x and y. It is an extension of Figure 7.1 in that it shows a background or framework for the setup in Figure 7.1. In this way, the two figures show explicitly the need for the value field for computation output and experimental outcome numbers. The figures also apply to numbers, in general, vectors, and other types of mathematical elements that include numbers in their description, not just to computer output and experiment outcome numbers.

8.2 Extended global mathematics

The essential role of the α field in assigning meaning to numbers and other mathematical elements is not limited to local mathematics with structures at each point of the manifold, M. It also applies to a global mathematics where mathematical systems of different types exist outside of space–time.

A good way to illustrate the importance of the α field to global mathematics is to consider extensions of the usual representation of mathematical systems. In the usual representations of global mathematics, there is just one structure, $\bar{S}$, for each system type, S. For global mathematics, these structures have no space or space–time location. For local mathematics, there is just one structure $\bar{S}_x$ at each location of a manifold, M. The extensions expand the number of structures of each type to be an infinite number of structures with different value factors.

The usual and extended global mathematics can be represented by unions over all structures that use numbers as part of their description. The usual global mathematics can be represented by the set, H^r, where

$$H^r = \cup_S \bar{S}^r. \tag{8.2}$$

The factor r is the same for all structures and is independent of location.

The extended global mathematics is represented by the expanded set:

$$G = \cup_r (\cup_S \bar{S}^r). \tag{8.3}$$

The union over S includes the system types that include numbers in their axiomatic description. The union over r for each system type S consists of the systems of each type with all possible positive real scaling factors, r. The only difference between this and the extended local mathematics of Eq. (8.1) is the lack of a location subscript.

In mathematics as usually practiced, the superscript, r, in Eq. (8.2) is set equal to 1. Here the description of the usual global mathematics is extended to an arbitrary value of r. The reason is that there does not seem to be any way to determine the value of r. Every mathematical element in a system of type S in H^r has a unique value. A number has a unique value. A vector has a unique value. For example, the number -4.12 has the value $v_r(-4.12)$ in the number structure, $\bar{S}^r$. Here v_r is the valuation function introduced in Section 1.3.

For the usual global mathematics, variations in the α field do not play a role. Numbers and their values, and vectors and the values of their scalar properties are the same at all locations. Space or space–time locations cannot be given to mathematical structures in H^r because they have none.

8.2.1 *The α field and extended global mathematics*

Extended global mathematics is similar to local mathematics in that the value field is essential. It is essential for the same reason in that it provides meaning or value to mathematical elements. Without α, the value or meaning of a number at a location in M is not known. It can belong to any one of an infinite number of number structures, each with different value factors. The value of the number depends on the value factor for the number structure containing the number.

For vector spaces, the values of scalar properties of a vector are not known. The values of these properties depend on the value factor for the scalar structure containing these scalars as numbers. Without the value factor, the values of these scalar properties cannot be determined.

This situation is illustrated in Figure 8.3 for computer output and experiment outcome numbers at locations, x and y. The collection of global number structures for the number type, S, is denoted by $\cup_r \bar{S}^r$. The figure shows that this collection is the same for both output and outcome numbers. The arrows represent two of the infinite number of possible choices for the structures containing the numbers.

The difference in the values of the output number for the two choices of value factors for the structures containing the number is easy to demonstrate. Let the output number d in $\bar{S}^t$ and $\bar{S}^s$ be given by

$$d = f_t = h_s. \tag{8.4}$$

Here f and h denote the values of d in $\bar{S}^t$ and $\bar{S}^s$.

The relation between the number values, f and h, is determined by the connection that maps the numbers in $\bar{S}^r$ to numbers in $\bar{S}^s$. This is given by

$$h = \frac{t}{s} f.$$

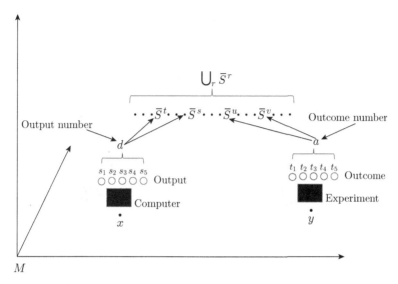

Figure 8.3. Schematic representation of computation output and experimental outcome states at locations x and y. The numbers represented by these states are shown as a and d. The global mathematical collection, $\cup_r \bar{S}^r$, showing an infinite number of structures of each type, is indicated by the horizontal dots. Four structures with value factors, t, s, u, v are shown. Two of the possible infinite number of number structures to which the numbers a and d can belong are indicated by arrows. The values or meanings of the numbers, a and d, depend on the value factor of the structure containing them.

A similar relation holds for the values of the outcome number a in $\bar{S}^u$ and $\bar{S}^v$. From

$$a = b_u = c_v, \tag{8.5}$$

one obtains

$$c = \frac{u}{v}b.$$

The presence of the α field removes this ambiguity in values of numbers for extended global mathematics. This results from the fact that the α field determines the value factors for numbers at each location in the manifold, M. This is shown in Figure 8.4 for the same computer output and experiment outcome numbers as in the previous figure. The collection of type S structures for all value factors shown in the figure remains the same.

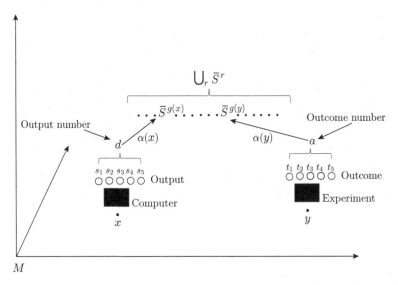

Figure 8.4. Use in global mathematics of the value field in selecting the number structure with the right scale factor to contain numbers at x and y. The structure type S is the same as that of the output and outcome numbers. The collection of type S number structures for all scaling factors is denoted by $\cup_r \bar{S}^r$.

The description of parallel transport of numbers and other types of mathematical elements in global mathematics is the same as that for local mathematics. Parallel transport of a number or other type of element in $\bar{S}^{g(y)}$ to the same element in $\bar{S}^{g(x)}$ is implemented by the same form of the connection. If $a_{g(y)}$ is contained in $\bar{S}^{g(y)}$, the same element in $\bar{S}^{g(x)}$ is defined by the connection $C_g(x,y)$ as in

$$a_{g(y)} \to C_g(x,y)a_{g(y)} = \left[\frac{g(y)}{g(x)}a\right]_{g(x)}. \tag{8.6}$$

The effect of the presence of α for the extended global mathematics shows that the usual global mathematics is a special case of the extended description. This can be seen by setting $\alpha(x) = \ln(r)$ for all x. For this case, with $g(x) = r$, one obtains the result that

$$G^g = H^r \tag{8.7}$$

with H^r given in Eq. (8.2). Also G^g is the restriction of G in Eq. (8.3) to one value of r, where $g(x) = r$ everywhere.

One sees from the above that the presence of a constant α field is implicit in the definition of the usual global mathematics. The value field has been hiding in plain sight. The discussion also shows that global mathematics can be extended to illustrate the implicit presence of the value field. It also shows that the effects and restrictions on variations of the value field described in previous chapters for local mathematics also apply for global mathematics. This includes the effects in gauge theories, quantum mechanics, and geometry.

8.2.2 Local and global mathematics

From the discussion presented so far, one might think that it would be sufficient to work just with extended global mathematics. Local mathematics would not be used, even in the presence of the value field. This is not the case.

To see this let x and y be points in M, where $g(y) \neq g(x)$. From Figure 8.4, one sees that $\bar{S}^{g(y)}$ is different from $\bar{S}^{g(x)}$. The value of number in one structure differs from that in the other structure. Local mathematics is implied because the structure $\bar{S}^{g(y)}$ is associated with location y and $\bar{S}^{g(x)}$ is associated with location x. The two structures cannot be the same because this would imply that a number has two different values in the same structure. This is impossible.

If $g(y) = g(x)$ for two points, x and y in M, then the structure $\bar{S}_x^{g(x)}$ is the same as the structure, $\bar{S}_y^{g(y)}$. The connection in Eq. (8.6) that maps $\bar{S}_y^{g(y)}$ onto $\bar{S}_x^{g(x)}$ is the identity. In this case, the structures can be considered either as local or global, at least for the two points, y and x.

If α is constant over some restricted region, Q, of M, then the local mathematical structures at all points in Q can be identified. The resulting structure, $\bar{S}_Q^k$, is global in Q. The structures at the points of Q can also be considered to be local. There is no difference between the two representations, as a collection of local structures in Q or as one structures global over Q.

Assume that Q is the region of space–time occupiable by us and includes nearby locations for which two-way communication with beings on other planets is possible. Then observers in Q will represent mathematics as global in the usual unextended sense. The mathematics used in theoretical descriptions of outcomes of experiments is either global or local. There would be no way to distinguish which mathematics one is using. Local mathematics would not be different from global mathematics.

In Chapter 6, it was seen that values of α at locations far outside of Q can be used to explain the observed red shift due to Hubble expansion and dark energy. Assume that this α-based description of the red shift for distant physical systems holds up and the Hubble expansion is valid everywhere. In this case, α is not constant anywhere, even in the region occupiable by us. This would be the case even if the variations in the value field would be too small to affect experimental properties of systems in the region, Q.

In this case, it would follow that mathematics is necessarily local. However, if the variation of the value field over Q is too small to affect theory predictions and their comparison with experimental outcomes, then local and global mathematics can be identified. The difference between assuming the structures are the same over Q or different would be too small to have any practical affect. This identification is limited to regions of space–time where variations in the value field are very small. In regions where this is not the case, mathematics is local.

8.3 Consequences of the nonexistence of the α field

In this work so far, the existence of the value field was implicitly assumed. It is worth exploring the consequences of physics and other sciences of the nonexistence of the value field. A good way to accomplish this is to examine the effect on the use of theory-experiment comparisons in the development of physics and other science.

One begins with the observation that agreement or disagreement between theory and experiment involves comparison of theory computer output number values with experiment outcome number values. Both theory computations and experiments give numbers as outputs. Without the presence of the α field, one cannot know the values of these numbers. Comparison of the number values between theory and experiment is not possible. This can be seen from the mathematical background for these numbers.

An output or outcome number at location x is a number in $\cup_r(S^r), x$ for global mathematics or $\cup_r(S_x^r), x$ for local mathematics. Here S is the number type appropriate for the output and outcome numbers and x is the location of the computer output or experiment outcome. From this, one knows that the number must have some value. However, the presence of the value field is essential for an observer to know the values of these numbers. Without the field present, an observer cannot know the values of the numbers. Comparison between theory and experiment becomes impossible.

These observations show that the lack of a value field, with extended global or local mathematics as a background, has a profound effect on physics and other sciences. A computer makes a good example. Justification that a computer does what it is intended to do is based on many layers of theory supported by experiment. These layers include electromagnetic theory, chemistry, quantum mechanics, special relativity, etc. Without this support, a computer is just a fancy contraption that does all sorts of weird things.

A similar argument applies to the apparatuses used in experiments. These are physical systems used to prepare a physical system in a desired state and to measure a property of the system in the state so prepared. Justification for the apparatuses doing what they are intended for is also based on many layers of theory supported by experiment. Without this justification, experimental equipment is a collection of physical systems with no meaning or use. They are contraptions.

The Large Hadron Collider is an example. Intrinsically, it is a meaningless physical system. Justification for the assertion that it is a machine that uses magnetic and electric fields to accelerate particles to very high energies and contain them is based on the validity of quantum electrodynamics (QED), special relativity, etc. The validity of these theories is based on many experiments that affirm the correctness of theory predictions.

An astronomical telescope is another example. Justification for the fact that the telescope does what it is supposed to do and is not a meaningless piece of machinery is based on the validity of optical theory and whatever other theories are used in the design and construction. Support for these theories is in turn based on experimental support by use of equipment and calculations.

The justification that the computers used to calculate predictions of theories, such as QED and special relativity, and equipment used for experiments that support these theories also depends on the assumption that the equipment for these predictions and experiment outcomes does what it is designed to do. The same argument for justification that the computer and equipment carry out their intended tasks for these experiments depends on prior theory predictions also supported by experiments.

This ordered sequence of justifications of experiments and calculations for each experiment reflects the growth of science in that it is rather like a tree. Twig tips represent a particular determination of experiment support or refutation of a theory. Downward motion from twigs to branches, stems, and trunk represents the earlier more elementary and crude science until one reaches the tree base. This corresponds to elementary sensations that are what we see, hear, feel, smell, and taste. Science began with the most elementary explanations of these sensations. Many steps in the development of science consisted of determination of the validity of theories by experiment support or refutation.

The point of this diversion into the brief description of the development of science is to point out that, for many steps, advancement depends on the comparison of theory prediction number values with experiment number values. The presence of the value or meaning field guarantees that the output and outcome numbers have values for comparison. Without the value field, α, comparison of theory computation output with experiment outcome becomes meaningless. The values of the computation output numbers and experiment outcome numbers are not known. Any values are possible. The whole tree edifice representing the progress of science would collapse. Physics and other quantitative sciences would not exist.

8.3.1   *Physical representations of numbers*

The point of this section is to emphasize the fact that representations of numbers are necessarily physical systems with no intrinsic meaning or value. Experiment outcomes and computer outputs are physical systems in different physical states. The form of the states of the physical systems determines their recognition as numbers. Examples include strings of symbols of a certain form written on paper, sound wave trains corresponding to spoken descriptions, or as some type of optical representation. These are all physical systems in some specified states. From their physical form, they are recognizable as numbers. That is, they have meaning as numbers. However, in the absence of the value field, their meanings or values are unknown.

Physical representations of mathematical elements and their properties are restricted to contain a finite amount of information. Representations of these elements are limited to those expressed by symbol strings of finite length. Elements requiring symbol strings of infinite length are excluded. It follows that natural numbers, integers, and rational numbers, both real and complex, are included. Almost all real and complex numbers are excluded because their representations require infinite length symbol strings. These numbers do not have a value because they do not have a physical representation.

The domain of α is not limited to physical representations of specific numbers. It also includes variable symbols for mathematical elements of different types. These include variable for numbers, vectors, operators, matrices, and many other types of elements. For this work, the domain of α is limited to the types of mathematical elements whose structures include numbers of some type in their axiomatic description.

The distinction between physical representations of mathematical elements and their properties and the meanings of these representations is not new. It plays an essential role in mathematical logic [14,22] where the distinction between syntactic and semantic structures is emphasized. Syntactic structures consist of many physical systems in different states. These are usually represented as finite length strings of symbols of different types. The symbol strings as physical systems in different states have no intrinsic meaning. The form and ordering of the systems show that they are strings of symbols. Strings of certain types are connected by logical rules of deduction. Theorems are symbol strings that appear anywhere in a sequence of applications of the deduction rules that begin from the initial axiom strings. None of the symbol strings have a meaning. Construction of theorems and other activities is like a game.

If the axioms are consistent, then the syntactic structure has associated one or more semantic structures or models. In these structures, symbol strings acquire meaning or value. For the different types of numbers and many other structures, there is just one (standard) model. These are the structures in general use. In these structures, numbers acquire the usual value. The rational number string 3.42 has the meaning 3.42 in the model of rational numbers.

The point to be emphasized here is that the existence of a semantic model of a consistent syntactic structure presupposes the existence of the value or meaning field. Without the presence of the value field, syntactic elements would just be physical systems in physical states. They would have no intrinsic meaning. Different types of symbol strings would be recognizable as numbers, proofs, theorems, etc. but they would not have meaning or value as numbers or their properties.

The situation described here is similar to the use of coding or encryption of messages. One is presented with the encrypted message in some physical form. But the meaning of the message is completely unknown. Decoding the message to give it meaning requires the presence of a key for the decoding process. Here numbers and other mathematical elements represented in some physical form are the coded expression. Decoding the numbers to determine their value or meaning requires the presence of a key. This is provided by the value field α.

The coding key provided by α is different from the usual coding key in that the coding depends on the space and time location of a string of symbols that is the physical representation of a number. The meaning or value of a number depends on its location. The location-dependent change in meaning or value of a number is determined by the space and time variation of the value field, α.

The similarity to coding emphasizes the fact that meaning and physics are closely related. A coded message is a physical system. It usually consists of an ordered collection of physical systems each of which are in certain physical states. Decoding is the process of determining the meaning of the collection of physical systems.

The same considerations apply to symbol strings that are text in a book. These have no intrinsic meaning. They represent a coded message. The decoding of these systems occurs quickly without much thought. But the meaning of these systems is also separate and distinct from the systems. The process of learning to determine the meaning of these systems is called learning to read.

One concludes from this discussion that the fact that numbers and other mathematical elements as physical systems in different states have meaning or value is *prima facie* evidence for the existence of a value or meaning field, α. This is the case whether α is a constant everywhere or varies with space or space–time.

Another way to express this is that the existence of the usual mathematics, global or local, is evidence for the existence of a value field. This field selects mathematical structures of different types with α determined scale factors from the extended local or global mathematics. As noted earlier without this selection, numbers and other mathematical elements would have no meaning or value.

It is possible that the same conclusions can be extended to the syntactical symbol strings representing formulas, theorems, and proofs that describe properties of numbers. As physical systems in different states, these symbol strings have no intrinsic meaning. The fact that they have meaning or value may also be due to the presence of a value or meaning field. This would be an extension of α from giving meaning to numbers and other mathematical elements to assign meaning or value to the syntactic symbol strings that describe properties of numbers and other mathematical elements in mathematical logic.

It is even possible that the domain of α can be extended to text in general. This book is an example. The observation that it has meaning might be evidence for the existence of an extended meaning or value field.

8.4   The α field and consciousness

So far it has been argued that experiment outcomes and theory computations, as physical systems in specific states, have no intrinsic meaning or value. The physical form of the states of these physical systems makes then recognizable as numbers of some type. However, without the presence of the α field, their value or meaning is unknown. This is also the case for any physical representations of numbers irrespective of their origin or the type of representation. The form of the physical representations makes them recognizable as numbers, but, in the absence of α, their meaning or value is unknown.

The same argument applies to vectors. A physical representation of a vector is a physical system in a state recognizable as a vector. Without the α field, the length value of the vector and the values of other numerical properties of the vector are not known.

One may argue against this position by noting that a conscious being can think about numbers and their properties. Thinking about numbers and vectors, and other elements and their properties is equivalent to thinking about the values or meanings of numbers, and vectors and other mathematical elements and their properties. These thoughts can occur with no sensory input of seeing or otherwise experiencing physical systems, such as symbol strings. These thoughts can occur to a person in a completely dark and silent room with no sensory input.

This argument is based on the observation that meaning or value is an essential property of consciousness. Without consciousness, nothing has meaning or value. It is an empty concept.

For us as conscious beings, the brain is the carrier of consciousness. It then seems reasonable to assume that different thoughts correspond to different physical states of the brain. For example, the appearance of the number value 3.752 in ones thoughts is expected to correspond to the change in the state of the brain from a state corresponding to no thought of the number value to a state corresponding to the thought of the number value.

At this point, very little is known about how the brain works. This is an area of much interesting and active research. The literature on the topic is large. This includes work on the biology and neural circuits in the brain and the relations between physics and consciousness [92]. Nevertheless, it is safe to assume that there is some type of correspondence between physical states of the brain and thoughts.

From a physical viewpoint, it is safe to assume the brain is a physical system that can be in a great number of complicated physical states. Extension of the train of reasoning about physical systems outside and external to the brain to the brain leads to the conclusion that the different physical brain states have no intrinsic meaning. If neural circuitry is an appropriate model for the brain, then the different brain states would correspond to different circuits being active. These circuits would have no intrinsic meaning. They would have no correspondence to thoughts or consciousness. In fact, consciousness would not exist.

One may speculate that the value field α can be extended to assign meaning or value to the brain states that correspond to numbers and properties of numbers. In this case, a conscious observer whose brain is one of these states would be thinking about, or be aware, of the number and its meaning or value. Without the α field, the observer may be aware of the number but would have no idea of the value or meaning of the number. It would be like a number expressed in a language with a different alphabet. If this is the case, then the presence of the value field, α, suitably extended, would act like an interpreter or decoder of at least some brain states. Symbols recognizable as numbers, basic operations on numbers and other properties, would have meaning or value to the observer.

This even suggests that the presence of the α field may be related to consciousness itself. The relation is based on the possibility that the value field is needed to determine the meaning associated with different physical states of the brain. If this is the case, and this is a big, speculative if, it would show that the presence of the value field is even more fundamental than has been described in this book. Whether this is the case or not must await much more work.

How the brain works is a fascinating subject. It may be hoped that the material described in this book will lead to a deeper understanding of the relationship between mathematics and the sciences and, possibly, consciousness. It is also hoped that this work will be of much help in constructing a coherent theory of physics and mathematics together.

Afterword

> The theorems are, certainly, very different
> from those to which we are accustomed,
> and they disconcert us a little at first.
>
> M. H. Poincaré[1]

The approach proposed by Paul Benioff should be seen in a broader context of mathematical modeling based on generalized arithmetic, a new scientific paradigm where arithmetic and calculus are rather loosely defined. From reading this book, we know that it is possible to formulate the laws of physics in terms of numbers that do not possess unique "numerical values". Somewhat surprisingly, it is possible to perform physics calculations even without prior specification of what should be exactly meant by "plus" or "times". It is enough to assume that multiplication and addition are commutative and associative, while multiplication is distributive with respect to addition. The resulting derivative is linear and satisfies the Leibniz rule. The associated integral is related to the derivative by the two

---

[1]M. H. Poincaré, "Les géométries non euclidiennes", *Revue Générale des Sciences Pures et Appliquées*, vol. 2, 769–774, 1891. The translated version appeared in *Nature*, vol. 45, 404–407, 1892.

fundamental theorems of calculus. Of course, the notion of linearity depends on what we mean by plus and times.[2]

Real numbers are here represented by elements of any set $\mathbb{X}$ whose cardinality equals that of the continuum. Such sets can be quite strange and conceptually very far from the naive picture of a "real line". A relation between $\mathbb{X}$ and physics is established at the very end, for example, by comparison with experiment. Arithmetic becomes in the new paradigm as physical as geometry in Einstein's general relativity. What is essential, $\mathbb{X}$ can be bijectively mapped onto $\mathbb{R}$ by means of some map $f_{\mathbb{X}} : \mathbb{X} \to \mathbb{R}$. The mapping is one-to-one but otherwise in principle arbitrary. This one-to-one relation is guaranteed by equal cardinalities of $\mathbb{X}$ and $\mathbb{R}$. The map "turns" elements of $\mathbb{X}$ into real numbers, simultaneously inducing in $\mathbb{X}$ the four basic arithmetic operations (addition, subtraction, multiplication, and division):

$$x_1 \oplus_{\mathbb{X}} x_2 = f_{\mathbb{X}}^{-1}\big(f_{\mathbb{X}}(x_1) + f_{\mathbb{X}}(x_2)\big), \tag{1}$$

$$x_1 \ominus_{\mathbb{X}} x_2 = f_{\mathbb{X}}^{-1}\big(f_{\mathbb{X}}(x_1) - f_{\mathbb{X}}(x_2)\big), \tag{2}$$

$$x_1 \odot_{\mathbb{X}} x_2 = f_{\mathbb{X}}^{-1}\big(f_{\mathbb{X}}(x_1) \cdot f_{\mathbb{X}}(x_2)\big), \tag{3}$$

$$x_1 \oslash_{\mathbb{X}} x_2 = f_{\mathbb{X}}^{-1}\big(f_{\mathbb{X}}(x_1)/f_{\mathbb{X}}(x_2)\big). \tag{4}$$

It also orders $\mathbb{X}$ ($x_1 \leq_{\mathbb{X}} x_2$ if and only if $f_{\mathbb{X}}(x_1) \leq f_{\mathbb{X}}(x_2)$). The new arithmetic, $\{\mathbb{X}, \oplus_{\mathbb{X}}, \ominus_{\mathbb{X}}, \odot_{\mathbb{X}}, \oslash_{\mathbb{X}}, \leq_{\mathbb{X}}\}$, is isomorphic to the standard arithmetic $\{\mathbb{R}, +, -, \cdot, /, \leq\}$. However, "isomorphic" is not synonymous to "physically equivalent". A generalized arithmetic implies a generalized (non-Newtonian) calculus which is isomorphic to the standard (Newtonian) calculus we learn at schools. As the readers have probably guessed, $f_{\mathbb{X}}$ has many analogies to Benioff's value maps.

A change of paradigm can be fruitful.

For example, in models of accelerated expansion of our universe, $\mathbb{X}$ is identified with time but the notion of dark energy is eliminated.

---

[2]In his remarks after Eq. (1.99), Paul Benioff stresses the role of linearity of his value maps and mentions the possibility of nonlinear generalization. However, all these generalized approaches are also based on linear maps, only the notion of linearity is formulated with respect to a more general arithmetic.

It is replaced by a nontrivial arithmetic of time. Comparison with experiment shows that

$$f_{\mathbb{X}}(t) = \frac{2}{3\sqrt{\Omega_\Lambda}} \sinh \frac{3\sqrt{\Omega_\Lambda}}{2} t \tag{5}$$

$$\approx 0.8 \sinh(t/0.8), \tag{6}$$

where t is computed since the Big Bang in units of the Hubble time. Expansion of the universe becomes non-Newtonian in two different meanings of the term — as non-Newtonian physics formulated by means of non-Newtonian mathematics [99].

The best known application of a similar "non-Diophantine arithmetic[3] philosophy" in a more standard context is encountered in $(1 + 1)$-dimensional special relativity. Elements of $\mathbb{X}$ then represent dimensionless velocities, $v/c \in \mathbb{X} = (-1, 1)$, and $f_{\mathbb{X}} = \tanh^{-1}$. If we extend the arithmetic interpretation to general relativity, the sets $\mathbb{X}$ have to be treated as fibers over some manifold. The velocities then parameterize Lorentz transformations in spaces tangent to the space–time manifold, so the arithmetic is defined locally at x by v_x/c_x, in close analogy to Benioff's value fields.

Although nothing can prevent us from writing $1 + 1 = 2$, relativistic nature performs the same calculation differently: $1 \oplus 1 = 1$, where

$$(v_1/c) \oplus (v_2/c) = \tanh \left(\tanh^{-1}(v_1/c) + \tanh^{-1}(v_2/c) \right) \tag{7}$$

$$= \frac{(v_1/c) + (v_2/c)}{1 + (v_1/c)(v_2/c)}. \tag{8}$$

As special relativity eliminated the luminiferous aether, the arithmetic implied by (6) seems to eliminate the dark energy. It is intriguing that in both cases we encounter hyperbolic functions. Perhaps one should take a closer look at this coincidence.

---

[3]Diophantus of Alexandria formalized arithmetic. Euclid of Alexandria formalized geometry. Mark Burgin formalized non-Diophantine arithmetic. We published a book whose first 600 pages, written by Mark, constitute the first monograph on non-Diophantine arithmetic — the field he created and seemed to consider his life achievement. The idea of getting me involved in editing Benioff's book is also due to Mark. He suddenly passed away on February 18, 2023.

Still, the addition of velocities is not accompanied in standard theory of relativity by an analogous multiplication:

$$(v_1/c) \odot (v_2/c) = \tanh\left(\tanh^{-1}(v_1/c) \cdot \tanh^{-1}(v_2/c)\right). \qquad (9)$$

Einstein's theory of relativity is thus formulated as if we already knew addition and subtraction but haven't heard of multiplication and division yet. Strictly speaking, the relativistic unit of velocity is not c but $c\,1_{\mathbb{X}} = c\tanh(1) = 0.76\,c$. I'm aware of only one place in relativistic physics where $0.76\,c$ plays a distinguished role [97].

In the context of Bell's theorem, $\mathbb{X}$ plays a role of quantum mechanical hidden variables [98–103]. Standard limitations imposed by Bell inequalities turn out to be circumvented by nontrivial additivity properties of non-Newtonian integrals. In this generalized world, Bell's theorem is untrue and quantum cryptography is not absolutely secure.

In psychophysical applications, $\mathbb{X}$ may represent states of a sensory system [61]. The bijection $f_{\mathbb{X}}$ is then responsible for logarithmic scales of decibels and star magnitudes (the Weber–Fechner law [104]). Ironically, 21st-century scientists behave analogously to their 19th-century colleagues who experienced gravity but searched in vain for some observable manifestations of non-Euclidean geometry of the universe. A physical non-Diophantine arithmetic is literally hiding just before our noses: Human and animal sensory systems perform a non-Diophantine subtraction.

In fractal applications, $\mathbb{X}$ represents Cantor dusts [56,93,94], Sierpiński carpets [95], or Koch curves [96]. Certain impossibility theorems about Fourier transforms on Cantor sets are easily circumvented by properties of non-Newtonian integration. The same happens with the difficulties that had plagued the problem of wave propagation on Koch curves.

The new framework is incredibly flexible and efficient from the point of view of mathematical modeling. The possibilities it opens are comparable to those created by non-Euclidean revolution in geometry. A relatively up-to-date account of the formalism can be found on Michael Grossman's website devoted to non-Newtonian calculus.[4] Non-Newtonian calculus as a tool for unification is described in the review [100].

---

[4]https://google.com/site/nonnewtoniancalculus/Home.

However, when we look from this perspective at the work of Paul Benioff, we notice a shift in emphasis. Benioff is not interested in just another mathematical tool. He wants to find a common ground for physics and mathematics. He is a little bit like Albert Einstein who searched for laws that determine both physics and geometry. For Einstein, geometry was not a tool for modeling physics — it was a subject of physics.

Benioff, in one of his earliest publications, contemplated the example of natural numbers represented solely by numbers that are even: 0, 2, 4, etc. From our arithmetic perspective, we would write $\mathbb{X} = \{0, 2, 4, \ldots\}$ and $f_{\mathbb{X}}(n) = n/2$. Multiplication and addition would read as follows:

$$n_1 \oplus_{\mathbb{X}} n_2 = f_{\mathbb{X}}^{-1}\big(f_{\mathbb{X}}(n_1) + f_{\mathbb{X}}(n_2)\big) = 2(n_1/2 + n_2/2) = n_1 + n_2, \tag{10}$$

$$n_1 \odot_{\mathbb{X}} n_2 = f_{\mathbb{X}}^{-1}\big(f_{\mathbb{X}}(n_1) \cdot f_{\mathbb{X}}(n_2)\big) = 2\big((n_1/2) \cdot (n_2/2)\big) = n_1 \cdot n_2/2. \tag{11}$$

As we can see, the addition is unchanged, but the multiplication is modified (this would be true for any linear $f_{\mathbb{X}}$). In particular, "one" is the element $1_{\mathbb{X}}$ satisfying

$$n \odot_{\mathbb{X}} 1_{\mathbb{X}} = n, \tag{12}$$

for any n. Clearly, $1_{\mathbb{X}} = f_{\mathbb{X}}^{-1}(1) = 2$. In this concrete example, the "value map" $f_{\mathbb{X}}(n) = n/2$ is linear and global. In Benioff's subsequent publications, the formalism evolved into its generalizations based on value maps which remained linear but became local — different at different space–time points, hence the name "value field".

My paths had crossed with Paul Benioff in 2014–2015 when I discovered the idea of a calculus based on generalized arithmetics. Actually, it later turned out that I only rediscovered (and somewhat generalized) an idea investigated many years earlier by Michael Grossman, Robert Katz, V. P. Maslov, or Endre Pap and, from a slightly different perspective, by Mark Burgin. However, in 2014, I was unaware of these works, but I vaguely remembered I had seen on arXiv.org some papers by Benioff on a similar topic. I emailed him, presenting a preliminary version of the preprint I was working on (and which finally evolved into [56]). We even discussed if the title should involve "relativity of arithmetic" or "relativity of arithmetics". I accepted Paul's suggestion to use the singular form "arithmetic", but apparently he had never read the published version of the paper, as he always quoted it with the original plural form of the early preprint. Neither I nor Paul were aware of the contributions by Grossman and Katz. The works by Endre Pap were brought to my attention by Peter Carr.[5]

We had never met in person, but we exchanged emails. Our paths had crossed and then parted — our scientific instincts and tastes were leading us in different directions. Paul was concentrated on the local aspect of the relation between physical reality and numbers. What I found more interesting was the generality inherent in nonlinear f_Xs. In this sense, there was absolutely no "conflict of interest" between me and Paul. We really had enough space to explore. Of course, we followed each other's works. I would not be surprised if he was a referee of some of my papers — I refereed some of his.

The adventure of editing his book was interesting and inspiring. It helped me to organize my thoughts, especially since I was writing my own "treaty against Bell" at the time [103]. I generally did not interfere with the text. I added a few references and a single footnote, sometimes I had to make a decision concerning an unfinished sentence or a formula that referred to some earlier version of the text. The draft was generally flawless, but I noticed an increasing number of typos as we approached the final chapters — as if the author was writing in a hurry.

Marek Czachor
Gdańsk University of Technology
25 April 2023

---

[5]Peter Carr was an outstanding financial scholar, the "C" in the CGMY model of pricing derivatives, and the dean of NYU Tandon Department of Finance and Risk Engineering. He died on March 1, 2022, exactly four weeks before Paul Benioff. I was shocked — we were in a middle of a discussion on his non-Newtonian reformulation of stochastic differential equations, when he stopped responding to my emails. I still have on my computer the draft of his last, unfinished paper "Lie's canonical coordinate and stochastic differential equations". His last published paper [105] refers to some of these ideas. It is a pity that our monograph [61] does not quote any work by Carr. Unfortunately, writing and editing of the book could not keep up with the constantly emerging new results.

References

[1] E. Wigner, "The unreasonable effectiveness of mathematics in the natural sciences", *Commun. Pure Appl. Math.*, **13**, No. 1, (1960). Reprinted in E. Wigner, *Symmetries and Reflections*, (Indiana Univ. Press, Bloomington, IN, 1966), pp. 222–237.

[2] R. Omnes, "Wigner's "Unreasonable effectiveness of mathematics", Revisited, *Found. Phys.*, **41**, 1729–1739, (2011).

[3] A. Plotnitsky, "On the reasonable and unreasonable effectiveness of mathematics in classical and quantum physics" *Found. Phys.*, **41**, 466–491, (2011).

[4] G. Boniolo, P. Budinich, and M. Trobok, "The role of mathematics in physical sciences", in G. Boniolo, P. Budinich, and M. Trobok, (eds.), Springer, Dordrecht, The Netherlands, 2005, Parts 2 and 3.

[5] Yu. I. Manin, "Mathematics and physics", in *Progress in Physics* vol. 3, A. Jaffe and Ruelle, (eds.), Birkhäuser, Boston, 1981 (Translated by Ann and Neal Koblitz). https://link.springer.com/chapter/10.1007/1-4020-3107-6_1.

[6] P. Benioff, "Towards a coherent theory of physics and mathematics", *Found. Phys.*, **32**, 989–1029 (2002); arXiv:quant-ph/0201093.

[7] P. Benioff, "Towards a coherent theory of physics and mathematics: The theory-experiment connection", *Found. Phys.*, **35**, 1825–1856, (2005).

[8] C. Pincock, "Proof and other dilemmas: Mathematics and philosophy", in B. Gold and R. Simons, (eds.), Spectrum Series, Mathematical Association of America, Washington DC, 2008, Chapter III. https://www.amazon.com/Proof-Other-Dilemmas-Mathematics-Philosophy/dp/0883855674/ref=monarch_sidesheet.

[9] J. Polkinghorne, (ed.), *Meaning in Mathematics*, Oxford University Press, Oxford, UK, 2011. https://www.amazon.com/Meaning-Mathematics-John-Polkinghorne/dp/019960505X.

[10] M. Tegmark, "The mathematical universe", *Found. Phys.*, **38**, 101–150 (2008).

[11] M. S. Leifer, Mathematics is physics, in Aguirre, A., Foster, B., Merali, Z. (eds.), Trick or Truth? *The Frontiers Collection.* Springer, Cham. https://doi.org/10.1007/978-3-319-27495-9_3.

[12] I. Volovich, "Number theory as the ultimate physical theory", *p-Adic Numbers Ultrametric Anal. Appl.*, **2**, No. 1, 77–87 (2010).

[13] S. Shapiro, "Mathematical objects", in *Proof and Other Dilemmas, Mathematics and Philosophy*, B. Gold and R. Simons, (eds.), Spectrum Series, Mathematical Association of America, Washington DC, 2008, Chapter III, pp. 157–178.

[14] H. J. Keisler, "Fundamentals of model theory", in *Handbook of Mathematical Logic*, J. Barwise, (ed.), North-Holland Publishing Co. New York, 1977, pp. 47–104.

[15] P. Benioff, "Relation between observers and effects of number valuation in science", *J. Cognitive Sci.*, **19**, No. 2, 229–251 (2018); arXiv:1804.04633.

[16] I. Montvay and G. Münster, *Quantum Fields on a Lattice*, Cambridge University Press, UK, 1994, Chapter 3.

[17] G. Mack, "Physical principles, geometrical aspects, and locality properties of gauge field theories", *Fortsh. Phys.*, **29**, 135 (1981).

[18] P. Benioff, "New gauge field from extension of space time parallel transport of vector spaces to the underlying number systems", *Int. J. Theor. Phys.*, **50**, 1887–1907 (2011); arXiv:1008.3134.

[19] P. Benioff, "Effects on quantum physics of the local availability of mathematics and space-time dependent scaling factors for number systems", in *Advances in Quantum Theory*, I. Cotaescu, (ed.), Intech 2012, Chapter 2; doi: 10.5772/36485. https://www.intechopen.com/chapters/28317.

[20] P. Benioff, "Gauge theory extension to include number scaling by boson field: Effects on some aspects of physics and geometry", in *Recent Developments in Bosons Research*, I. Tremblay, (ed.), Nova Publishing Co., 2013, Chapter 3; arXiv:1211.3381. https://www.amazon.com/Recent-Developments-Research-Physics-Technology/dp/1624179606.

[21] P. Benioff, "The no information at a distance principle and local mathematics: Some effects on physics and geometry", to appear in M. Burgin and G. Dodig-Crnkovic, (eds.) *Theoretical Information Studies*; arXiv:1803.00890.

[22] J. Barwise, "An introduction to first order logic", in *Handbook of Mathematical Logic*, J. Barwise, (ed.), North-Holland Publishing Co. New York, 1977. pp. 5–46.

[23] J. L. Bell, "From absolute to local mathematics", *Synthese*, **69**, 409–426 (1986).

[24] Y. Manin, *Gauge Field Theory and Complex Geometry*, 2nd edition, translated from Russian by N. Koblitz and J. King, Grundlehren der mathematischen wissenschaften, Springer Verlag, Berlin, 1997, pp. 289.

[25] C. N. Yang and R. L. Mills, "Conservation of isotopic spin and isotopic gauge invariance", *Phys. Rev.*, **96**, 191–195 (1954).

[26] P. Benioff, "Fiber bundle description of number scaling in gauge theory and geometry", *Quantum Stud. Math. Found.*, **2**, 289–313 (2015); arXiv:1412.1493.

[27] M. Daniel and G. Vialet, "The geometrical setting of gauge theories of the Yang-Mills type", *Rev. Modern Phys.*, **52**, 175–197 (1980).

[28] W. Drechsler and M. Mayer, *Fiber Bundle Techniques in Gauge Theories*, Springer Lecture Notes in Physics #67, Springer Verlag, Berlin, 1977.

[29] P. Benioff, "Effects of number scaling on entangled states in quantum mechanics", in *Quantum Information and Computation IX*, E. Donkor, M. Hayuk, (eds.), Proc. of SPIE conference, vol. 98730, SPIE, Bellingham, WA, 2016, arXiv:1603.04752.

[30] R. Kaye, *Models of Peano Arithmetic*, Clarendon Press, Oxford, UK, 1991, pp. 16–21.

[31] Wikipedia: Integer. https://en.wikipedia.org/wiki/Integer.

[32] A. Weir, *Lebesgue Integration and Measure*, Cambridge University Press, New York, 1973, p. 12.

[33] J. Randolph, *Basic Real and Abstract Analysis*, Academic Press, Inc. New York, 1968, p. 26.

[34] J. Shoenfield, *Mathematical Logic*, Addison-Wesley Publishing Co. Inc. Reading, MA, 1967, p. 86; Wikipedia: Complex Numbers.

[35] C. Pfeifer, "The tangent bundle exponential map and locally autoparallel coordinates for general connections with applications to Finslerian geometries", *Int. J. Geom. Methods Mod. Phys.*, **13**, No. 3, 1650023 (2016), arXiv:1406.5413. https://www.worldscientific.com/doi/abs/10.1142/S0219887816500237.

[36] P. Moylan, "Fiber bundles in nonrelativistic quantum mechanics", *Fortschr. Phys.*, **28**, 269–284 (1980).

[37] R. Sen and G. Sewell, "Fiber bundles in quantum physics", *J. Math. Phys.*, **43**, 1323–1339 (2002).

[38] H. J. Bernstein and A. V. Phillips, *Sci. Am.*, **245** No. 1, 122–137 (1981).

[39] B. Iliev, "Fiber bundle formulation of nonrelativistic quantum mechanics", *J. Phys. A: Math. Gen.*, **34**, No. 23, 4887 (2001). arXiv:quant-ph/0004041. https://iopscience.iop.org/article/10.1088/0305-4470/34/23/308.

[40]   M. Asorey, J. Carinena, and M. Paramio, "Quantum evolution as a parallel transport", *J. Math. Phys.*, **23**, No. 8, 1451–1458 (1982).

[41]   D. Husemöller, *Fibre Bundles*, 2nd edition, Graduate Texts in Mathematics, vol. 20, Springer Verlag, New York, 1975.

[42]   D. Husemöller, M. Joachim, B. Jurco, and M. Schottenloher, *Basic Bundle Theory and K-Cohomology Invariants*, Lecture Notes in Physics, Springer, Berlin, Heidelberg, 2008, p. 726. doi: 10.1007/ 978-3-540-74956-1, e-book.

[43]   Wikipedia: Fiber bundles. https://en.wikipedia.org/wiki/Fiber_ bundle.

[44]   C. F. Pewster, "Lectures on quantum field theory in curved spacetime", Lecture Note No. 39, Max Planck Institute, Leipzig, 2008.

[45]   T. P. Cheng and L. F. Li, *Gauge Theory of Elementary Particle Physics*, Oxford University Press, Oxford, UK, 1984, Chapter 8.

[46]   M. Peskin and D. Schroeder, *An Introduction to Quantum Field Theory*, Addison-Wesley Publishing Co., Reading, MA, 1950, Chapters 4 and 15. https://en.wikipedia.org/wiki/Fiber_bundle.

[47]   Y. Aharonov and D. Bohm, "Significance of electromagnetic potentials in the quantum theory", *Phys. Rev.*, **115**, 485 (1959). https:// journals.aps.org/pr/abstract/10.1103/PhysRev.115.485.

[48]   M. Czachor, "Dark energy as a manifestation of nontrivial arithmetic", *Int. J. Theor. Phys.*, **56**, 1364–1381 (2017); arXiv:1604.05738.

[49]   H. Weyl, "Gravitation and electricity", Sitsungsberichte der Koniglich Preussichen Akademie der Wissenschaften, January–June, pp. 465–480, 1918.

[50]   L. O'Raifeartaigh, *The Dawning of Gauge Theory*, Princeton Series in Physics, Princeton University Press, Princeton, NJ, 1997.

[51]   P. Benioff, "Effects of a scalar scaling field on quantum mechanics", *Quant. Inform. Process.*, **15**, No. 7, 3005–3034 (2016); arXiv: 1512.05669.

[52]   P. W. Higgs, "Broken symmetries and the masses of gauge bosons", *Phys. Rev. Lett.*, **13**, No. 16, 508, (1964).

[53]   Wikipedia: Atlas (topology). https://en.wikipedia.org/wiki/Atlas_ (topology).

[54]   P. Ginsparg, "Applied conformal field theory" in E. Brézin and J. Zinn–Justin, (ed.) *Ecole d'Eté de Physique Théorique: Champs, Cordes et Phénomènes Critiques/Fields, Strings and Critical Phenomena* (Les Houches, France), Session XLIX, Elsevier Science Publishers B.V. 1989; arXiv:hep-th/9108028. https://cds.cern.ch/ record/204595/.

[55]   M. Gaberdiel, "An introduction to conformal field theory", *Rept. Prog. Phys.*, **63**, 607–667 (2000); arXiv:hep-th/9910156.

[56] M. Czachor, "Relativity of arithmetic as a fundamental symmetry of physics", *Quantum Stud. Math. Found.*, **3**, 123–133 (2016); arXiv:1412.8583.

[57] M. S. Burgin (2010). "Nonclassical models of the natural numbers", *Gen. Math., Uspekhi Mat. Nauk*, **32**, 209–210 (1977), in Russian.

[58] M. Burgin, "Non-Diophantine arithmetics or is it possible that 2+2 is not equal to 4?" Ukrainian Academy of Information Sciences, Kiev, 1997 (in Russian); "How we count or is it possible that two times two is not equal to four", Elsevier, Preprint 0108003, 2001; Science Working Paper S1574-0358(04)70635-8, p. 12. https:// papers.ssrn.com/sol3/papers.cfm?abstract id=3153531, http://www. sciencedirect.com/preprintarchive.

[59] M. Burgin, "Diophantine and non-Diophantine arithmetics: Operations with numbers in science and everyday life", LANL, Preprint Mathematics GM/0108149, 2001, p. 27, arXiv:math/0108149.

[60] M. Burgin, "Introduction to projective arithmetics", *Gen. Math.*, (2010); arXiv:1010. 3287. https://arxiv.org/abs/1010.3287.

[61] M. Burgin and M. Czachor, *Non-Diophantine Arithmetics in Mathematics, Physics, and Psychology*, World Scientific, Singapore, 2020, doi: 10.1142/11665.

[62] M. Grossman and R. Katz, *Non-Newtonian Calculus*, Lee Press, Pigeon Cove, 1972.

[63] M. Grossman, *The First Nonlinear System of Differential and Integral Calculus*, Mathco, Rockport, 1979.

[64] M. Grossman, *Bigeometric Calculus: A System with Scale-Free Derivative*, Archimedes Foundation, Rockport, 1983.

[65] V. P. Maslov, "On a new superposition principle for optimization problems", *Uspekhi Mat. Nauk.*, **42**, 43 (1987), in Russian.

[66] V. N. Kolokoltsov and V. P. Maslov, *Idempotent Analysis and Applications*, Kluwer, Dordrecht, 1997.

[67] E. Pap, "g-calculus", *Zb. Rad. Prirod.–Mat. Fak. Ser. Mat.*, **23**, 145 (1993).

[68] E. Pap, "Generalized real analysis and its applications", *Int. J. Approx. Reason.*, **47**, 368–386 (2008).

[69] M. Grabisch, J.-L. Marichal, R. Mesiar, and E. Pap, *Aggregation Functions*, Cambridge University Press, Cambridge, 2009.

[70] Wikipedia: Curvature. https://en.wikipedia.org/wiki/Curvature.

[71] Wikipedia: Conformal map; see also Conformal manifolds. https:// en.wikipedia.org/wiki/Conformal_map.

[72] S. Carroll, "Lecture notes on general relativity", December, 1997, NSF-ITP/97-147; arXiv:gr-qc/971209. https://ned.ipac.caltech.edu/ level5/March01/Carroll3/Carroll_contents.html.

[73] Wikipedia: Comoving and proper distances. https://en.wikipedia. org/wiki/Comoving_and_proper_distances.

[74] Wikipedia: Red shift. https://en.wikipedia.org/wiki/Redshift.

[75] Wikipedia: Hubble's law. https://en.wikipedia.org/wiki/Hubble% 27s_law.

[76] Wikipedia: The expansion of the universe. https://en.wikipedia.org/ wiki/Expansion_of_the_universe.

[77] R. Durrer, "What do we really know about dark energy?" *Phil. Trans. Royal Soc. A*, **369**, 5102–5114 (2011); arXiv:1103.5331v3 [astro-ph.CO].

[78] Wikipedia: Recombination (Cosmology). https://en.wikipedia.org/ wiki/Recombination_(cosmology).

[79] A. D. Dolgov, "Neutrinos in cosmology", *Phys. Rep.*, **370**, 333–535 (2002); arXiv:hep-ph/0202122v2.

[80] Wikipedia: Search for extraterrestrial intelligence. https://en. wikipedia.org/wiki/Search_for_extraterrestrial_intelligence.

[81] H. A. Smith, "Alone in the universe", *Am. Sci.*, **99**, No. 4, 320 (2011).

[82] F. Bezrukov and M. Shaposhnikov, "The standard model Higgs boson as the inflaton", *Phys. Lett. B*, **659**, 703–706 (2008).

[83] I. Zlatev, L. Wang, and L. P. Steinhardt, "Quintessence, cosmic coincidence,and the cosmological constant", *Phys. Rev. Lett.*, **82**, No. 5, 896–899 (1999); arXiv:astro-ph/9807002.

[84] F. Asenjo and S. Hojman, "Class of exact solutions for a cosmological model of unified gravitational and quintessence fields", *Found. Phys.*, **47**, 887–896 (2017).

[85] R. Brandberger, R. R. Cuzinatto, J. Frölich, and R. Namba, "New scalar field quartessence", *JCAP*, **2**, No. 43 (2019), arXiv:1809.07409. https://iopscience.iop.org/article/10.1088/1475-7516/2019/02/043.

[86] A. H. Q. Peter, "Dark matter: A brief review", *Proceedings of Frank N. Bash Symposium 2011: New Horizons in Astronomy*, **149**, No. 14 (2012); arXiv:1201.3942. https://pos.sissa.it/149/014/.

[87] L. Randall, "Dark matter and the dinosaurs: The astounding inter-connectedness of the universe", Ecco/Harper Collins Publishers, New York, 2015, ISBN 978-0-06-232847-2.

[88] Miao Li, Xiao-Dong Li, Shuang Wang, and Yi Wang, "Dark energy, a brief review", *Front. Phys.*, **8**, 828–846 (2013).

[89] P. J. E. Peebles and B. Ratra, "The cosmological constant and dark energy", *Rev. Modern Phys.*, **75**, No. 2, 559–606 (2003); arXiv: astro-ph/0207347.

[90] G. K. Goswami, A. Pradhan, and A. Beesham, "A dark energy quintessence model of the universe", *Mod. Phys. Lett. A*, **35**, No. 04, 2050002 (2020). https://www.worldscientific.com/doi/abs/10.1142/S0217732320500029.

[91] M. Rinaldi, "Higgs dark energy", *Classic. Quant. Grav.*, **32**, 045002 (2015); arXiv:1404.0532v4.

[92] S. Gao, "A quantum theory of consciousness", *Minds Mach.*, **18**, 39–52 (2008).

[93] D. Aerts, M. Czachor, and M. Kuna, "Crystallization of space: Space-time fractals from fractal arithmetic", *Chaos Solitons Fractals*, **83**, 201 (2016).

[94] D. Aerts, M. Czachor, and M. Kuna, "Fourier transforms on Cantor sets: A study in non-Diophantine arithmetic and calculus", *Chaos Solitons Fractals*, **91**, 461 (2016).

[95] D. Aerts, M. Czachor, and M. Kuna, "Simple fractal calculus from fractal arithmetic", *Rep. Math. Phys.*, **81**, 357 (2018).

[96] M. Czachor, "Waves along fractal coastlines: From fractal arithmetic to wave equations", *Acta Phys. Polon. B*, **50**, 813 (2019).

[97] M. Czachor, "Cosmic-time quantum mechanics and the passage-of-time problem", *Universe*, **9**, 188 (2023).

[98] M. Czachor, "A loophole of all 'loophole-free' Bell type theorems", *Found. Sci.*, **25**, 971 (2020).

[99] M. Czachor, "Non-Newtonian mathematics instead of non-Newtonian physics: Dark matter and dark energy from a mismatch of arithmetics", *Found. Sci.*, **26**, 75–95 (2021); arXiv:1911.10903 [physics.gen-ph].

[100] M. Czachor, "Unifying aspects of generalized calculus", *Entropy*, **22**, 1180 (2020).

[101] M. Czachor, "Arithmetic loophole in Bell's theorem: Overlooked threat to entangled-state quantum cryptography", *Acta. Phys. Polon. A*, **139**, 70–83 (2021).

[102] M. Czachor and K. Nalikowski, "Imitating quantum probabilities: Beyond Bell's theorem and Tsirelson bounds", *Found. Sci.* (2023). https://link.springer.com/article/10.1007/s10699-022-09856-y.

[103] M. Czachor, "Contra Bellum: Bell's theorem as a confusion of languages", *Acta Phys. Polon. A*, **143**, No. 6, S158–S170 (2023). http://appol.ifpan.edu.pl/index.php/appa/article/view/143_s158/143_s158.

[104] J. C. Falmagne, *Elements of Psychophysical Theory*, Oxford University Press, Oxford, 1985.

[105] P. Carr and U. Cherubini, "Option pricing generators", *Front. Mathe. Finance*, **2**, No. 2, 150–169 (2023).

Index